International Conference on Science, Infrastructure Technology and Regional Development (ICoSITeR 2017)

Energy Security for Enhancing National Competitiveness

IOP Conference Series: Earth and Environmental Science
Volume 124

South Lampung, Indonesia
25 – 26 August 2017

ISBN: 978-1-5108-5921-0
ISSN: 1755-1307

Printed from e-media with permission by:

Curran Associates, Inc.
57 Morehouse Lane
Red Hook, NY 12571

Some format issues inherent in the e-media version may also appear in this print version.

Printed by Curran Associates, Inc. (2018)

For permission requests, please contact the Institute of Physics
at the address below.

Institute of Physics
Dirac House, Temple Back
Bristol BS1 6BE UK

Phone: 44 1 17 929 7481
Fax: 44 1 17 920 0979

techtracking@iop.org

Additional copies of this publication are available from:

Curran Associates, Inc.
57 Morehouse Lane
Red Hook, NY 12571 USA
Phone: 845-758-0400
Fax: 845-758-2633
Email: curran@proceedings.com
Web: www.proceedings.com

Table of contents

Volume 124

International Conference on Science, Infrastructure Technology and Regional Development (ICoSITeR) 2017 "Energy Security for Enhancing National Competitiveness"

25–26 August 2017, South Lampung, Indonesia

Accepted papers received: 16 February 2018
Published online: 5 March 2018

Preface

Papers

IOP Conf. Series: Earth and Environmental Science **124** (2018) 011001 doi:10.1088/1755-1315/124/1/011001

WELCOME SPEECH FROM CONVENER OF ICOSITER 2017

Dear ladies and gentlemen,

Tabik Pun,

On behalf of organizing committee of ICoSITeR 2017, I am pleased to welcome you to the 2nd International Conference on Regional Science, Infrastructure Technology and Regional Development hosted by Institut Teknologi Sumatera. This valuable international conference will be held on 25 to 27 August 2017 which will closed by a field trip program.

This Conference is an event series of Dies Natalis of ITERA. The Conference is presenting five invited speakers for two days. They are 1) Arcandra Tahar, Ph.D., our Vice Minister of Energy and Mineral Resources who represented to Prahoro Yulijanto Nurtjahyo, Ph.D.; 2) Prof. Doyoung Byun from Sungkyunkwan University, South Korea; 3) Prof. Fujii, from Gunma University, Japan; 4) Prof. Toto Winata from Institut Teknologi Bandung, and 5) Dr. Rahayu Sulistyorini, from Institut Teknologi Sumatera.

This year, ICoSITeR 2017 has a big theme: "**Energy Security for Enhancing National Competitiveness**". Growing interest in industrial era, energy security has been becoming the most-spoken discussion topic both in Indonesia and other countries around the world. In Indonesia, the energy crisis caused insecurity to reliability of fuel supply in meeting the needs.

Through this conference, we would like to engage all stakeholders consist of academia, business sectors, government and community in an open and scientific discussion about the problems and potentials of our future energy so that our national competitiveness can be enhanced in the future.

The first day of ICoSITeR starts with an opening ceremony followed by a plenary session presenting two invited speakers Prahoro Yulijanto Nurtjahyo, Ph.D from Ministry of Energy and Mineral Resources and Prof. Fujii, from Gunma University, Japan. After the plenary session, the discussion will be divided into several parallel classes which has a various open and constructive dialogues and presentations from our valuable presenters. We also have a Gala Dinner presenting special performances.

The second day of ICoSITeR, starts with a plenary session presenting Prof. Doyoung Byun from Sungkyunkwan University, South Korea. For the next session, we have parallel session which is divided into four parallel classes as in Conference day-1. As the parallel session finished, the Conference presents Prof. Toto Winata from Institut Teknologi Bandung, and Dr. Rahayu Sulistyorini from Institut Teknologi Sumatera in the last plenary session.

The interest of ICoSITeR is clear. The Conference has attracted 142 participants. We will hear 87 oral presentations. The subjects range from 1) New trends and technology for energy security, 2) Policies and strategies for energy security, 3) Energy conversion studies, 4) Decision support system for energy security, 5) Artificial intelligence and machine learning studies for energy security 6) Computational methods for energy security, 7) Energy savings for urban livelihood, 8) Public awareness and education for energy security, 9) Renewable energy systems in smart cities, 10) Future challenges of energy security. For publication purposes, we offer to our presenter to submit their paper to IOP Proceeding, indexed by Scopus. As the closing of the Conference series, we also offer a field trip program that will trip to Pahawang Island on 27 August 2017.

Finally, I would like to deliver my appreciation for all parties for having a good cooperation, strong effort and coordination for making successfully ICoSITeR 2017. My special thanks to our rector, Prof. Ofyar Z. Tamin who fully supports ICoSITeR 2017, also Prof. Mitra Djamal and Prof. Sukrasno, I thank you for your supports. I should express my gratitude to all sponsors who support the success of ICoSITeR 2017.

Concluding, I wish you in the name of ICoSITeR 2017 an enjoyable and enlightening conference.

Thank you!

Convener of ICoSITeR 2017

Prof. Leo H. Wiryanto

IOP Conf. Series: Earth and Environmental Science **124** (2018) 011001 doi:10.1088/1755-1315/124/1/011001

Advisory Board

Prof., Ir. Ofyar Z. Tamin, M.Sc (Eng)., Ph.D – *Rector of Institut Teknologi Sumatera, Indonesia*

Steering Committee

Prof. Dr.Ing Mitra Djamal –*Vice Rector, Institut Teknologi Sumatera, Indonesia*

Prof. Dr. Sukrasno, M.S. – *Vice Rector, Institut Teknologi Sumatera, Indonesia*

Organizing Committee

Prof. Dr. Leo Hari Wiryanto, M.S. – *Conference Chair*

Dr. Ir. Citra Persada, M.Sc. – *Conference Co-Chair*

Helmia Adita Fitra S.T., M.T. – *Secretary*

Restu Almiati S.Si., M.T. – *Treasurer*

Hadi Teguh Yudistira S.T., Ph.D. – *Sponsorship and Conference Management System*

Muhammad Asril, M.Si. – *Conference Management System*

Rishal Asri, S.T., M.Eng. – *Publication Information and Field Trip*

Kardo Rajagukguk, S.Pd., M.Eng. – *Publication Information and Public Relation*

Efa Maydhona Saputra, M.T. – *Logistics and Appurtenances*

IOP Conf. Series: Earth and Environmental Science **124** (2018) 011001 doi:10.1088/1755-1315/124/1/011001

Office

Telp	: (0721) 8030188, (0721) 8030189
Email	: icositer@itera.ac.id
Website	: www.icositer.itera.ac.id/ www.itera.ac.id
Contact Person	: Mr. Rishal : +62 852 5580 7138
	Mr. Kardo : +62 813 6111 3856

Venue

Institut Teknologi Sumatera

Jalan Terusan Ryacudu, Way Hui, Kecamatan Jati Agung

Lampung Selatan, Provinsi Lampung, Indonesia 35365

IOP Conf. Series: Earth and Environmental Science **124** (2018) 011002 doi:10.1088/1755-1315/124/1/011002

Peer review statement

All papers published in this volume of *IOP Conference Series: Earth and Environmental Science* have been peer reviewed through processes administered by the proceedings Editors. Reviews were conducted by expert referees to the professional and scientific standards expected of a proceedings journal published by IOP Publishing.

IOP Conf. Series: Earth and Environmental Science **124** (2018) 012001 doi:10.1088/1755-1315/124/1/012001

An Analysis of the Influence of Variability Rainfall on Flow Rate Based on the Watershed Characteristics

B A Sulistyono[1], L H Wiryanto[2]

Department of Mathematics Education, PGRI Nusantara University Kediri, Indonesia
Department of Mathematics, Bandung Institute of Technology, Indonesia

[1]bb7agus1@gmail.com, [2]leo@math.itb.ac.id

Abstract. Flood is one of natural phenomena that often occurs during the rainy season. In general, flooding occurs due to the inability of the river to accommodate the discharge that enters into it. The rain that descends in a watershed will turn into a stream. Thus, there is a relationship between rainfall and flow rate (discharge), which depends on the geometric conditions of watershed. The aim of this research is to determine the amount of discharge that outflow from the watershed due to the rain that descends in the area. The process of transformation rainfall into discharge is done by using Nakayasu synthetic unit hydrograph method. Derivation of this method is done by using different rainfall input i.e. single station and rainfall watershed. The results unit hydrographs are then used for design flood calculation in river.

1. Introduction

Flood is a natural phenomena that often occurs during the rainy season. In general, flooding occurs due to the inability of the river to accommodate the discharge that enters into it. The river's ability to accommodate discharge is called the river capacity. The magnitude of the river capacity is determined by the area of the wet cross section and the flow velocity. While the flow velocity depends on the basic slope of the river, the roughness of the river body, and the hydraulic radius. In short, the river capacity depends on the shape of the river geometry. The rain that descends in a watershed will caus the river discharge to increase. If the discharge coming into the river exceeds the river capacity, there will be overflow from the river.the overflow of this river is one of the causes of flooding in a region. The relationship between rainfall and discharge, which depends on the geometry of the watershed, is interesting to study especially assoiated the amount of discharge that outflow from watershed.

Research on the transformation of rainfall into discharge and its methods has been done by several researcher, among them, Sihotang, et al. (2011) conducted a study to calculate the design flood discharge at Gintung Dam using synthetic unit hydrograph of Nakayasu method with rainfall as its input. The calculation results show that the maximum discharge that enters the dam due to rain for return period 100 years is 224,401 m^3/s. Hasibuan, S.H. (2012) conducted a study to analyze flood design discharge at Bonai River, Kunto Darussalam Residence, Rokan Hulu District. Method was used to analyze is synthetic unit hydrograph of Nakayasu by rainfall in Pasar Kampar Station at year 2000 to 2009. The result of the research prove that flood design discharge at Bonai River for return period 0 years is 286,75 m^3/s. Suhariyanto, S (2014) analyzes the flow discharge occurring as a due to of falling rainfall in the Kali Sampean river basin using synthetic unit hydrograph of Nakayasu. The results showed that the rainfall that caused floods in river Kali Sampean in 2002 and 2008, is a rainfall with return period 100 years and produce flow discharge of 1.833,594 m^3/s.

Different from Sihotang, et al. in this study, rainfall data has been obtained from the rainstation for transformed into a discharge without having to select that the maximum rainfall data. The transformation process of rainfall into discharge will be done by using Nakayasu synthetic unit hydrograph. In Indonesia especially in Java Island this method is often used for discharge analysis. This is because the number and distribution of rainfall observation station is still lacking. The expected result of this research is to know the amount of discharge out of the watershed outlet due to input variation.

2. Theorical basic

In planning in the field of water resources, it is often necessary that the flood discharge data plan realistic. Flood plans with certain repeat periods can be calculated and flooding data or rainfall data. If flood discharge data is available long enough (more than 20 years), the flood discharge can be directly calculated with method of probability analysis. Medium if the data available only in the form of rain data and watershed characteristics , one of the recommended methods is calculating flood discharge from the maximum daily rainfall data plan with the superposition of unit hydrograph.

2.1. Watershed Characteristisc

Watershed or drainage basin is an area on the surface of the earth in which there is a drainage system consisting of one main stream and some of his tributaries, which function as a catchment area and steam water through one outlet. Several characteristics of watersheds that are important to examine as follows.

Drainge area. The drainage area is the most important watershed characteristic in watershed modelling. The drainage area reflect the volume of water that can result from falling rainfall in the area. Uniform or uniform rainfall for the entire drainage area is a common assumption in hydrological modelling.

Watershed length. The length of watershed is usually defines as the distance measured along the main river from the outlet to the watershed. The river usually will not reach the watershed limit, so it needs to be drawn the extension line starting from the edge of the river the the watershed limit by taking into account the flow drawn. Although the drainage are and the length of watershed are the size of the watershed but they reflect different aspects of the size. The drainage area is used as an indication of the potential for rain to produce a certain amount of water volume, while the length of the watershed is usually used in the calculation of the travel time required by water to flow within the watershed.

Watershed slope. Flood is a quantity that reflects the momentum of runoff and slope is an important factor in the momentum. The slope of the watershed reflects the rate of change of elevation within a certain distance along the main flow direction. The slope is measured by the elevation difference between the two main river ends divided by the length of the watershed. Different elevations do not always be or reflect the maximum elevation difference in the watershed. The highest elevation is usually located along the watershed and the edge of the river or main stream generally reach the watershed.

2.2. Unit Hydrograph (UH)

The flow hydrograph represents a time distributionof the flow (in this case the discharge) in the river in a watershed at a given location. A river flow hydrograph is an essential part of any water resources field. There is a close relationship between the hydrograph with the characteristic of a watershed, where the the flood hydrograph can show the watershed response to the rain input.

The unit hydrograph is direct runoff hydrograph produced by effective rainfall that occurs evenly across the watershed and with fixed intensity for a set time unit, that called the unit rain. The unit rain is a long rainfall such that the duration of runoff does not become short even though the rainfall becomes short. Thus the rain of the selected unit is the same length or shorter than the hydrograph rising period (time from the starting point of the surface stream to the peak). The runoff period from the unit rain ia all about the same and has nothing to do with the intensity of the rain.

The concept of unit hydrograph is based on linear systems theory and follow the principles of superposition and proportionality. For example, if one inch of excess rainfall produces a direct runoff peak of 100 cfs then two inch of excess rainfall with produce a direct runoff of 2 x 100 =200 cfs. Similarly if one inch of rainfall is followed by two inch of rainfall, the hydrographs from both rainfall pulses are simply added after accounting for the necessary time lag. The following figure demonstrates this concept.

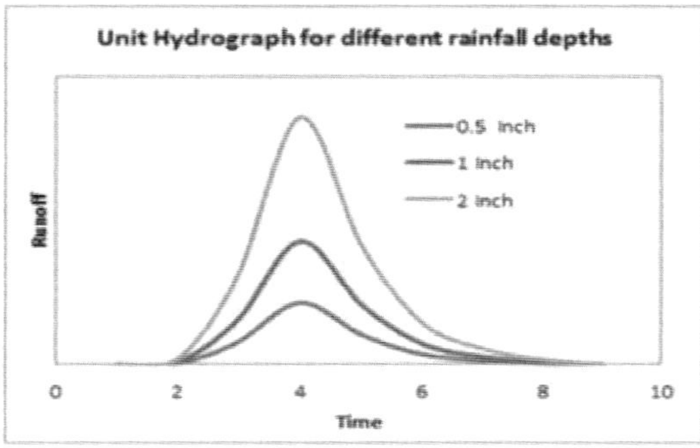

Figure 1. Unit hydrograph for different rainfall depth

Assuming that the watershed response to rain is linear, it is less price. Nevertheless, the use of unit hydrograph has yielded many satisfactory results for various condition. Thus, the theory of unit hydrograph is widely used in determining discharge or flood plans.

2.3. Synthetic Unit Hydrograph of Nakayasu

As indicate earlier, an effective collection of effective rainfall observation and direct runoff is required to derivation the unit hydrograph. Thus, the resultant of the unit hydrograph is specific to a particular watershed that determined by the point in the stream where the direct runoff observation is performed. If no direct observation is available, if no a unit hydrograph for another location at a stream in the same watershed or for a nearby catchment with similar charactristics is required, a synthetic unit hydrograph procedureshall be used.

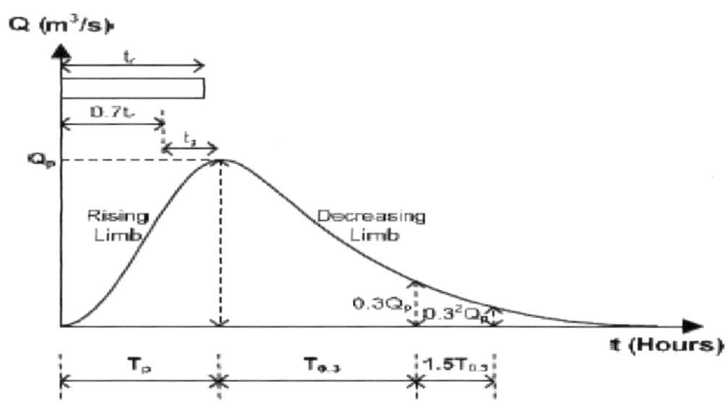

Figure 2. Scheme of Synthetic Unit Hydrograph of Nakayasu

Nakayasu has provided a synthetic unit hydrograph based on his research in Japan, as shown in Fig. 2. There are four variables have to be obtained, i.e. time lag (T_p), peak discharge (Q_p), rising limb curve (Q_a), and decreasing limb curve (Q_d). each variable will be explained as follows : time lag is the function of the time of concentration (t_g) and the duration of the effective rainfall (t_r). the equation of time lag is follows

$$T_p = t_g + 0.8\,t \tag{1}$$

Peak discharge is determined by some factors such as watershed area (A), watershed characteristic coefficient (C), unit rainfall (R_o), time lag, and time required to discharge reduction up to 30% peak discharge ($T_{0.3}$). The relationship among the variables is shown as follows

$$Q_p = \frac{C.A.R_0}{3.6(0.3T_p + T_{0.3})} \tag{2}$$

The rising limb curve is the function of time lag, peak discharge, and time. The function of Q_a is expressed as follows

$$Q_a = Q_p\left(\frac{t}{T_p}\right)^{2.4} \tag{3}$$

The decreasing limb curve is influenced by time lag, peak discharge, and time required to discharge reduction up to 30% of the peak discharge. Its function is follows

$$Q_d = Q_p(0.3)^{\frac{(t-T_p)}{T_{0.3}}} \quad , Q_d > 0.3Q_p \tag{4}$$

$$Q_d = Q_p(0.3)^{\frac{(t-T_p)+0.5T_{0.3}}{1.5T_{0.3}}} \quad , 0.3Q_p > Q_d > 0.3^2Q_p \tag{5}$$

$$Q_d = Q_p(0.3)^{\frac{(t-T_p)+1.5T_{0.3}}{2T_{0.3}}} \quad , 0.3^2Q_p > Q_d \tag{6}$$

3. Methodology

The synthetic unit hydrograph parameter values can be analysed with various supporting data. The coefficient of surface flow depends on the data of land use, the area of the watershed, the length of the river flow is analysed from the topographic map. After the unit hydrograph can be calculated, the analysed the flow discharge data by using the proposed proposition of hydrograph is a constan base width argument, linearity proposition, and superposition proposition. For hydrograph analysis required is data in the form annual rain.

In general, available rainfall data in the form of daily rainfall data. In order to be used in the formula Nakayasu synthetic unit hydrograph, then the daily rainfall data needs to be converted into hour time rainfall data. In order to obtain carefulresults, the rain stations used to analyse the rain data are selected stations that are in the watershed and that is spreaded evenly. The rain data used is the rainfall data for the last 10 years. Of available daily rain data, one daily rainfall data is selected. Thus, within one year there will be 12 daily rainfall data.

The next step is to test consistency, calculate the everage rainfall by arithmetic method or Thiessen polygon. Then, see the appropriate distribution model, whether the data follows the normal distribution, log normal, log Piersen III, or Gumbel. After finding the appropriate distribution model, then calculated the rainfall design for certain return period. To convert daily data into hourly rainfall is used Mononobe method. Whereas Mononobe equation is as follows

$$I = \frac{R_{24}}{24}\left[\frac{24}{t_c}\right]^{2/3} \tag{7}$$

where I is rainfall intensity, R_{24} is daily rainfall, and t_c is time concentration or rain duration.

4. Results

The watershed data used in this research is secondary data [3]. The data is the result of measurement and recording obtained from various related agencies. Existing data include 1:50.000 scale topographic map, and data of discharge measurement, rainfall data,from six watershed in South Sulawesi. Based on data collected conducted data selection for further analysis and from topography map determined parameters of each watershed. From the topography map, the parameters of the watershed are determined as followed, after the unit hydrograph has been obtained, adjusted the unit hydrograph volume so that the value is equal to the runoff height of 1 milimeter on entire the watershed. The result obtained is a unit hydrograph representing each watershed as presented in Fig. 3

for the Jenalata watershed, whereas the parameters of the hydrograph forming for each watershed is presented in Table 1.

Table 1. Parameters of watersheds

No	Watershed Parameters	Names of Watershed					
		Jenelata	Noling	Kalaena	Maros	Walanae	Maloso
1.	Watershed Area (A)	227.6	595.5	933.3	274.2	2354	1476
2.	Watershed Area Upstream	117.13	297.7	422.27	135.79	1233.28	581.00
3.	River Slope	0.0227	0.00072	0.00194	0.00289	0.00068	0.00158
4.	River Length Main (L)	36.6	68.3	74.875	38	175	106
5.	Time rises hydrogrph (T_r)	2.45	2.68	3.05	2.28	24.07	3.82
6.	Peak Discharge (Q_p)	9.24	23.45	33.20	12.22	32.83	44.90
7.	Base time Hydrograph (T_b)	23.82	32.63	29.69	27.84	45.56	29.88

Based on the recorded discharge data obtained from each selected flood case and then sought the unit hydrograph with the steps as described previously. Of each flood case after its unit hydrograph is obtained, then averaged to obtain the measured unit hydrograph of each watershed reviewed as presented in Table 2.

Table 2. The main parameters of measured unit hydrograph

No	Watershed Names	Year of data recording until 1985			Year of data recording until 1999		
		Q_p	T_p	T_b	Q_p	T_p	T_b
1.	Jenelata	10.93	2.00	19.00	8.75	2.22	25.11
2.	Noling	32.77	2.22	22.85	30.75	2.00	22.58
3.	Kalaena	38.73	3.75	23.75	45.59	1.50	17.50
4.	Maros	13.31	3.57	25.86	-	-	-
5.	Walanae	-	-	-	-	-	-
6.	Maloso	-	-	-	54.86	3.60	23.20

When considered the performance of measure unit hydrograph compared to the Nakayasu synthetic unit hydrograph show a relatively large deviation, see Table 3.

Table 3. Value of measured unit hydrograph parameter and Nakayasu and its deviation

Parameters of hydrograph		Names of Watershed					
		Jenelata	Noling	Kalaena	Maros	Walanae	Maloso
Peak Discharge	measurable Nakayasu	10.93 10.1	32.77 15.29	38.73 22.04	13.31 11.79	- 24.99	- 25.25
	Deviation (%)	-7.59	-53.22	43.09	-11.42	-	-
Peak Time	measurable Nakayasu	2.00 1.89	2.22 3.27	3.75 3.55	2.57 1.95	- 7.91	- 4.91
	Deviation (%)	-5.5	47.29	-5.3	24.12	-	-

Thus, in order to for Nakayasu syntethic unit hydrograph usage to be optimized in South Sulawesi, the factors causing deviation on flood peak discharge are examined more closely. By not ignoring the deviations in the use of Nakayasu synthetic unit hydrograph, the use of the method of analysis with this model is considered very advantangeous to be applied by adjusting the parameters of forming peak discharge, so the use of this model by itself will reduce the difficulty of determining the discharge, especially in rivers that have not available discharge pauge at the control point.

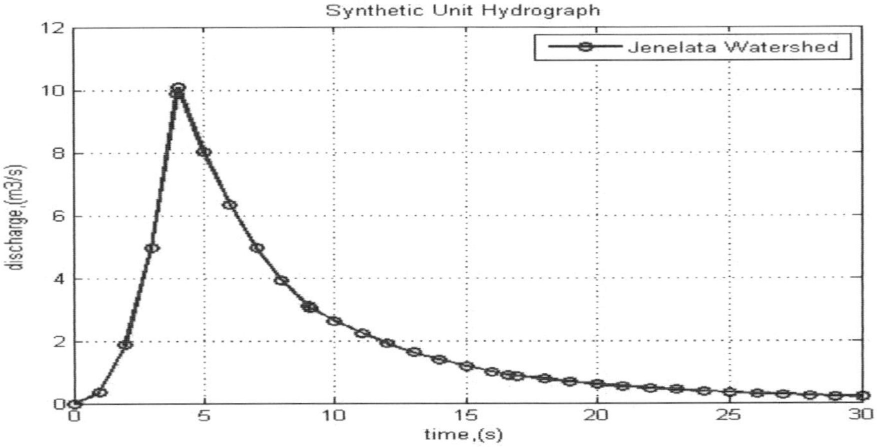

Figure 3. Nakayasu synthetic unit hydrograph of Jenelata Watershed

5. Resumes

The results obtained based on data analysis before 1985 compared with last year's data analysis result showed a small significant difference during that period the watershed did not change significantly. The most sensitive watershed physical factor to the hydrograph peak is the area of the watershed, while the length of the main river is very sensitive to the peak time. Of several watershed in South Sulawesi tested with Nakayasu, the results show considerable deviations to measurable hydrograph for each watershed observed, especially at their peak discharge values due to the uneven distribution of rainfall stations in each watershed.

References

[1] Hasibuan, S H 2011 Analisa debit banjir sungai Bonai Kabupaten rokan Hulu menggunakan pendekatan hidrograf satuan Nakayasu, *Jurnal APTEK*, 4(1), 23-28.

[2] Kusumastuti, D I dan Jokowinarno, D. 2012 Time step issue in unit hydrograph for improving runoff prediction in small catchment, *Journal of Water Resource and Protection*, 4, 686-693.

[3] Maricar, F dan Sri Harto, Br. 2001 Analisa kepekaan hidrograf satuan sintetik Gama I dalam penentuan debit banjir rancangan, *FORUM TEKNIK*, 25(8), 315-328.

[4] Safarina, A B 2012 Modified Nakayasu synthetic unit hydrograph method for meso scale ungange watersheds, *International Journal of Engineering Research and Application* (IJERA), 2(4), 649-654.

[5] Salim, N 2015 Determining the coefficient of T, A, and river length of flood runoff model using synthetic unit hydrograph Nakayasu, *IOSR Journal of Mechanical and Civil Engineering* (IOSR-JMCE), 12(6), 24-31.

[6] Sharma, R K 1980 Hydrology and water resourses, *Dhanpat Rai & Son*, New Delhi India.

[7] Sosrodarsono, S dan Takeda, K 1983 Hidrologi untuk pengairan, *Pradnya Paramita*, Jakarta.

[8] Sri Harto, Br 1993 Analisis hidrologi, *Gramedia Pustaka Utama*, Jakarta.

[9] Subarkah, I 1980 Hidrologi untuk perencanaan bangunan air, *Penerbit Idea Darma*, Bandung.

[10] Suhariyanto, A 2014 Prediksi titik-titik banjir berdasarkan kondisi geometri sungai, *Jurnal Rekayasa Sipil*, 8(3), 229-238.

[11] Triadmodjo, B 2009 Hidrologi terapan, *Beta Offset*, Yogyakarta.

Analysis on Transportation Infrastructure Availability to Achieve Environmental and Social Sustainability in Karawang

A D Rarasati[1,2] and N B Octoria[1,2]

[1] Civil Engineering Department, Faculty of Engineering, Universitas Indonesia, Kampus UI Depok 16424, Indonesia
[2] Center for Sustainable Infrastructure Development, Faculty of Engineering, Universitas Indonesia, Kampus UI Depok 16424, Indonesia

E-mail: ayomi@eng.ui.ac.id

Abstract. Sustainable infrastructure is the key to development success. At the same time, transportation infrastructure development will involve social and environmental conditions of the local surroundings. Assessment of the availability of such transport infrastructure is one of the solutions adapted from social and environmental impacts. By conducting a correlation test, the presence of transportation infrastructure and the social conditions of the environment can be identified. The results obtained show that the accessibility, the level of security, and the level of equality are correlated to social and environmental sustainability in Karawang. In terms of environment, the availability of transportation infrastructure is not directly related to the impact of environmental sustainability. The impact of the perceived environment also has no effect on the journey. Correlation results indicate that the length of travel time and congestion level do not make the perceived impact greater. The impact of the perceived environment is merely due to the high utilization of private vehicles in Karawang which subsequently leads to higher energy consumption.

1. Introduction

Development process will bring social impacts to the targeted community. These impacts have the potential to create social issues stemming from the direction of the relevant development policies, most likely because these policies were established by the government to focus its orientation on growth or physical development. In the meantime, the government is not giving adequate attention to existing social and cultural conditions. The government as the determinant of development policy often sets ideal images from its own perception, without inviting public contribution in shaping these images. The same thing applies in transportation infrastructure development which has not put into consideration the social needs in the targeted areas.

One of the said targeted areas for transportation infrastructure development is the Regency of Karawang, West Java Province, Indonesia. According to the Regency Planning and Development Board in 2016, Karawang is strategic because it is part of Pantura route, Jakarta – Bandung toll roads, and the intersection of 3 National Activity Center (Bodebek – Bandung – Cirebon). Such strategic location makes Karawang ideal for development projects, especially considering the plan to build Jakarta – Bandung fast train with Karawang as one of its transit stations. Therefore, existing transportation infrastructure must be well analyzed in order to ensure sustainability.

IOP Conf. Series: Earth and Environmental Science **124** (2018) 012002 doi:10.1088/1755-1315/124/1/012002

In addition to social sustainability, transportation infrastructure planning as well as any development policies must also have environmental sustainability. Consequently, this research is aimed at analyzing the current existing transportation infrastructure in Karawang for that purpose.

2. Literature review

Sustainable infrastructure transportation is a necessary approach in creating physical facilities to support both present and future economic, social and environmental objectives. For this purpose, the Indonesian National Planning and Development Agency established 9 indicators, namely movement activities, air pollution, road safety, economic productivity, accessibility, economic change, changes in land usage, asset ownership, and transportation policy and planning. These indicators are similar to those stated by Kumar, et al. [1], who grouped them into 3 categories: social, environmental and economic indicators.

According to Tumlin [2], land transportation consists of many aspects such as roads for bicycles, pedestrians, and other motor vehicles, and railways for trains, LRT, etc. Nonetheless, sustainable transportation indicators themselves cannot be defined directly due to different standards applied in each region with different stakeholders.

With regards to the environmental aspect, a few things must be controlled to ensure environmental sustainability: fuel consumption, pollution level, construction materials such as asphalt, and aggregates. This consideration also applies to pedestrian and bicycle facilities within public transportation system [3].

In terms of social aspect, a sustainable transportation system is needed to ensure every community is connected and their interests are promoted. It can also be seen from road accessibility and availability for pedestrians and bicycles, two things which also contribute to communication between communities. Their fuel consumption and resulting pollution will bring direct impacts to the surrounding environment.

3. Research methodology

The research methodology consists of research instruments in the form of questionnaires designed to assess social impacts which will be influenced by local accessibility and the ensuing infrastructure development. Data sampling was carried out in two stages: pilot survey and main survey. The pilot survey was conducted for ±4 days to test the questionnaires' readability. Afterwards, the main survey was conducted for 3 weeks in Karawang by taking respondent samples in a random manner from the streets and door-to-door. The areas covered were Teluk Jambe Timur, Karawang Barat, Kertasari, and Batujaya with a total of 344 respondents.

There are 5 variables related to social aspect, namely time, traffic congestion, transportation mode used, social welfare, and safety for road users. The variables related to the environmental aspect were land damage, air pollution, noise pollution and energy consumption.

These variables were the basis used to create the questionnaires in Likert scale, covering also origin and destination of respondents' daily activities to work/school, health centers, shopping centers, and worship centers. The questionnaire results correlated the Likert scale answers and the Origin-Destination answer. Spearman Test was selected to correlate the perceived conditions and the existing conditions, while Mann U Whitney and Scoring Tests were chosen to correlate other answers with the primary data. These test results both directly and indirectly described any existing relationships.

4. Results and Discussion

The respondents' demography consisted of 60% women and 40% men. In terms of age, the most respondents were within the age range of 17 – 25 years, with 42% of them students/college students and latest education at senior high or vocational school. Based on domiciles, the most respondents came from Teluk Jambe Timur since the survey was conducted mostly in urban areas.

4.1. Analysis on the Connection between Transportation Infrastructure Availability and Social Sustainability

This research assesses respondents' mobilization towards locations that can support social sustainability of the community such as education, work, health centers, shopping centers and worship facilities. Over 50% of these respondents used motor vehicles to reach said locations, except worship centers. For worship purposes, 74% of these respondents preferred to walk due to the near proximity of the worship centers.

Safety level is the only variable that highly influenced community's mobilization towards these locations, even though the correlation remains the same positive value, low to medium, in describing the correlations between travel time, congestion level, and the expenses for each journey towards those locations.

Community participation is a prerequisite in achieving social sustainability, as seen in rural and urban areas that are exposed to technology and outside influence from different communities. Karawang's participation level (with regards to safety facilities, reprimand against violations, assistance for the disable) is very high, with only a difference in that rural communities have more urge to assist the disable is using transportation facilities.

4.2. Analysis on the Connection between Transportation Infrastructure Availability and Environmental Sustainability

Further tests were performed to learn about the environmental sustainability in Karawang, as divided into 2 areas: rural and urban. This division was meant to identify the perceived environmental impacts on different levels of advancement in the transportation infrastructure as well as the population density in both areas.

Basically, all respondents perceived negative impacts due to transportation, with air pollution as the most perceived impact (75%). No correlation was found between the perceived impacts and the travel time and level of congestion. This showed that even though the travel time and congestion level are low, the respondents still perceived significant environmental impacts.

4.3. Accessibility and Mobility in Karawang

The minimum service standards on transportation infrastructure can be seen in the performance index, which depends on, among others, the accessibility to a certain area. This, in turn, depends on the length of existing roads. According Bappenas in 2012, the accessibility index is obtained by dividing the total length of roads with the width of the area in km^2. Data in 2016 showed that Karawang is 1,753.27 km^2 wide with a total road of 2,682.06 km in length [4]. Therefore, Karawang's accessibility index can be calculated as below:

$$\text{Accessibility Index: } \frac{\text{Length of Roads (km)}}{\text{Width (km}^2\text{)}}$$

$$\text{Accessibility Index : } \frac{2{,}682.06\ \text{km}}{1{,}753.27\ \text{km}^2} = 1.53\ \text{km/km}^2$$

According to The Ministry of Public Works Regulation No. 534/KPTS/M/2001, the accessibility index depends on population density. Different densities in different regions will have different needs for road networks. The higher the accessibility index of a region, the more effective its road networks are in servicing its community. Population density can be obtained by comparing the number of population to the area width. In 2016, Karawang had a population density of 1,094 people/km^2, a number which is considered high. However, its accessibility index was 1.53km/km^2, higher than the required minimum standard of 1.5 km/km^2.

In addition to the accessibility index, road performance can also be measured from the region's mobility level by calculating the total length of existing roads (km) per 1000 population. In 2016, Karawang had 2,273,579 populations and therefore its mobility can be calculated as below:

IOP Conf. Series: Earth and Environmental Science **124** (2018) 012002 doi:10.1088/1755-1315/124/1/012002

$$\text{Mobility Index} = \frac{\text{Length of Roads (km)}}{\text{1000 Population (people)}}$$

$$\text{Mobility Index} = \frac{\text{2,682.06 km}}{\text{2,273 people}} = \text{1.17 km/person}$$

The two road performance factors showed that Karawang has fulfilled the required length of roads; however, the quality of these roads tends to be poor as observed by the researchers and can be seen in Figure 1.

The 2016 BPS data on Karawang [4] stated that 39.37% of its regency and village roads were categorized as poor while 26.25% were categorized as satisfactory. Hence, to meet the minimum road standards, the regional administration should improve the quality of Karawang's roads.

Figure 1. Road condition in Karawang Regency

4.4. The Connection between Transportation Infrastructure Availability and Social Sustainability

Transportation infrastructure availability is used as a benchmark to achieve social and environmental sustainability since it is expected to support community's mobility. Therefore, this research is aimed at learning how the availability of transportation infrastructure influences the respondents' desire to travel. Social sustainability is achieved when community can fulfill its basic needs such as adequate education, health service, and income. Transportation infrastructure supports the community's needs to reach the locations of these basic needs.

The first significance level is in the mobility towards work areas. Basically, all social-based sustainable transportation variables influence community's mobility. The scoring test on safety, travel time and congestion level variables showed that these had the highest assessment weight. Duration and congestion level variables had higher correlation values compared to cost variable due to varieties in the cost required by workers to mobilize them to work. Everyone's standard is different so the amount spent for transportation to work has different significance between one worker and another. An example is by comparing farmers and office workers, where the same amount of transportation expenses might have higher significance to farmers' mobilization than that of office workers. Therefore, although transportation cost can influence mobilization, there is only little correlation between the amount and its significance. In addition, the travel time between one profession and another is different: even a little time is valuable for office workers since they do not wish to be late, whereas farmers and some other professions have a little more freedom in this regard. The average duration was 26.26 minutes with 15.1 minutes of congestion. In conclusion, duration has medium significance in influencing respondents' mobility.

Judging from the mapping, the respondents were mainly local farmers as well as workers in industrial areas in the regency capital city. The mobilization of workers towards the capital city and its surrounding areas was mostly due their employment in these industrial areas, especially in *Karawang*

International Industrial City (KIIC). This mobilization data is also supported by BPS, who stated that Karawang's workers mostly work in industry sector, even though some people domiciled farther away from the capital city chose to become local farmers. The ability of industry to absorb labor and the possible income it offers are attractive, therefore enticing people to work in the capital city and its surrounding regions. The influence of travel time and cost to mobility caused many workers to move and live closer to the capital city in order to save time and minimize transportation expense to work. In turn, this would increase the population in the capital city and its surrounding areas. To avoid this, it is therefore necessary to accommodate the workers' needs for shorter travel time and minimum transportation cost. At the same time, the development of effective and inexpensive transportation infrastructure must be accompanied by improvement in facilities for farmers, such as work loan. This would help towards avoiding the decrease in the number of farmers who might be tempted to leave farming in favor of industrial work. Therefore, the number of industrial workers and farmers would remain balanced.

In addition to employment, education quality also plays an important role to achieve social sustainability. Therefore, transportation infrastructure that supports the community to reach their education centers must be prioritized in details. The average travel time required to reach education centers is 27.79 minutes with 10 minutes' congestion. In general, students perceived travel time, congestion level and cost as having similar significance. All three influences their mobility to schools but the correlation tends to be weak.

Similar to workers, many Karawang students traveled towards the capital of Karawang and its surrounding areas, including those further away from the capital. This tendency is understandable since universities are located in these areas such as Universitas Singaperbangsa, the only state university which is located near the capital in Teluk Jambe Timur. The need for decent education is a basic right for the people of Karawang and the need to mobilize towards education centers must be fulfilled with effective transportation infrastructure. This improved connection between areas inside Karawang must also be accompanied with quality education infrastructure especially in its remote areas for the sake of education equality among these areas.

In addition to the needs to mobilize towards education centers and business centers, social sustainability is also achieved when community can fulfill their basic needs for food, health and worship. These three aspects similarly require transportation infrastructure to connect community to the desired locations for these purposes. Correlation and significance tests showed that in general all transportation factors namely accessibility, safety, and modes influence mobility although they may have weak correlation.

4.5. Connection between Transportation Infrastructure Availability and Environmental Sustainability

The impacts perceived by the community showed that some factors were not correlated to the environmental impacts they felt. The correlation test showed that the length of travel time and congestion level was not related to the impacts. All respondents feel that transportation brought negative impacts to Karawang's environment in the form of damage, air pollution and noise pollution. The use of private vehicles especially motorbikes was one of the main contributor to this. Table 1 shows the minimum number of people using public transportation.

The high percentage of private vehicles usage especially motorbike means higher fuel consumption for transportation in Karawang. According to 2015 data from the Directorate General of Land Transportation in, each motorbike consumed 1500 KJ of energy, higher than that of public transportation. This level of energy consumption presents one transportation problem for Karawang to achieve environmental sustainability.

Table 1. Transportation Modes Used to Reach Locations

Transportation Mode	Work	School	Market	Health Center	Worship Center
Motorbike	**54%**	**73%**	**80%**	**80%**	19%
Private car	18%	5%	9%	11%	3%
Public car	1%	3%	3%	2%	1%
On foot	25%	10%	5%	6%	**76%**
Rented Motorbike	1%	6%	4%	1%	1%
Bicycle	1%	1%	0%	0%	1%
Train	0%	1%	0%	0%	0%
Total	100%	100%	100%	100%	100%

5. Conclusion

Karawang has adequate quantity of road infrastructure for decent accessibility and mobility; however, 26.25% of these roads are categorized are poor quality. This research showed that there is a correlation between transportation infrastructure and social sustainability as seen from the community's mobility in fulfilling its basic needs. This mobility is highly influenced by the levels of accessibility and safety, two factors with the most significance.

With regards to environmental sustainability, Karawang has rather high energy consumption, environmental damage and pollution level. However, the perceived environmental impacts are not correlated with the perceived travel factors even though the environmental impacts stemmed from the high percentage of private vehicle usage, especially motorbike, and high fuel consumption. The high percentage of private vehicle usage itself is due to lack of decent public transportation options.

Transportation indicators such as accessibility, mobility, safety and equality in Karawang need to be considered in accordance with the community condition in order to achieve sustainability. In addition, environmental sustainability is low when environmental damage, air pollution and noise pollution is high and affecting community. Transportation infrastructure in Karawang can still achieve social and environmental sustainability but it is lacking in integrated public transportation modes and decent road condition.

Acknowledgement

The authors would like to gratitude Universitas Indonesia through PITTA Grant No. 743/UN2.R3.1/HKP.05.00/2017 for conducting this research and disseminating research results and the Center for Sustainable Infrastructure Development for the inkind contribution.

References

[1] Ravindra K, Errampali M, Amit D and Sanjeev S 2015 Analytical hirarchy process for assessing sustainability: indicators of public transportation systems, pedestrians and feeder services in developing countries *World Journal of Science, Technology and Sustainable Development* **12** pp 281-293.

[2] Jeffrey T 2012 *Sustainable Transportation Planning: Tools for Creating Vibrant, Healthy, and Resilient Communities* (New Jersey: John Wiley & Sons, Inc.)

[3] Dušan T, and Milan J 2016 *Transportation Engineering: Theory, Practice, Modeling* (Elsevier)

[4] Karawang Statistic Authority 2016 *Karawang dalam Angka 2016* (Karawang: Badan Pusat Statistik Kabupaten Karawang)

Application of Multipurpose Cadastre to Evaluate Energy Security of Land Parcel (Case Study: Gedung A and Gedung B, Institut Teknologi Sumatra)

S M Alif[1], A P Nugroho[1], B E Leksono[2]

[1] Geomatics Engineering, Institut Teknologi Sumatera, Lampung Selatan, Lampung, Indonesia
[2] Geodesy and Geomatics Engineering, Institut Teknologi Bandung, Bandung, Jawa Barat, Indonesia

Abstract. Energy security has one of its dimensions: Short-term energy security which focuses on the ability of the energy system to react promptly to sudden changes within the supply-demand balance. Non-energy components (such as land parcel) that comprise an energy system are analysed comprehensively with other component to measure energy security related to energy supply. Multipurpose cadastre which is an integrated land information system containing legal, physical, and cultural is used to evaluate energy (electrical energy) security of land parcel.

The fundamental component of multipurpose cadastre used to evaluate energy security is attribute data which is the value of land parcel facilities. Other fundamental components (geographic control data, base map data, cadastral data) are used as position information and provide weight in room (part of land parcel) valuation. High value-room means the room is comfortable and/or used productively by its occupant. The method of valuation is by comparing one facility to other facilities. Facilities included in room valuation are relatively static items (such as chair, desk, and cabinet) except lamps and other electronic devices. The room value and number of electronic devices which consume electrical energy are correlated with each other. Consumption of electrical energy of electronic devices in the room with average value remains constant while consumption in other room needs to be evaluated to save the energy.

The result of this research shows that room value correlate weakly with number of electronic device in corresponding room. It shows excess energy consumed in low-value room. Although numbers of electronic devices do not always mean the consumption of electrical energy and there are plenty electronic devices, it is recommended for occupant to be careful in utilizing electronic devices in low-value room to minimize energy consumption.

1. Introduction

Energy security which is the uninterrupted availability of energy sources at an affordable price has one of its dimensions: Short-term energy security which focuses on the ability of the energy system to react promptly to sudden changes within the supply-demand balance. Non-energy components (such as land parcel) that comprise an energy system are analysed comprehensively with other component to measure energy security related to energy supply. Multipurpose cadastre, which is an integrated land information system containing legal (e.g. property ownership or cadastre), physical (e.g. topography, man-made features), and cultural (e.g. land use, demographics), is used to evaluate energy (electrical

IOP Conf. Series: Earth and Environmental Science **124** (2018) 012003 doi:10.1088/1755-1315/124/1/012003

energy) security of land parcel. Multipurpose cadastre is ideal solution as inputs as well as instrument in establish modern land administration system [1]. Multipurpose cadastre can be used as basic information and backbone for government policy [2].

The fundamental component of multipurpose cadastre used to evaluate energy security is attribute data which is the value of land parcel facilities. Other fundamental components geographic control data, base map data, cadastral data) are used as position information and provide weight in room (part of land parcel) valuation [3]. High value-room means the room is comfortable and/or used productively by its occupant. The case studies of this research are Gedung A and Gedung B that are located in land parcel owned by Institut Teknologi Sumatra in Lampung Province, Indonesia. The location of case study is shown on **Figure 1.** This research will show the method to evaluate excess energy consumption of room based on productivity and electronic devices.

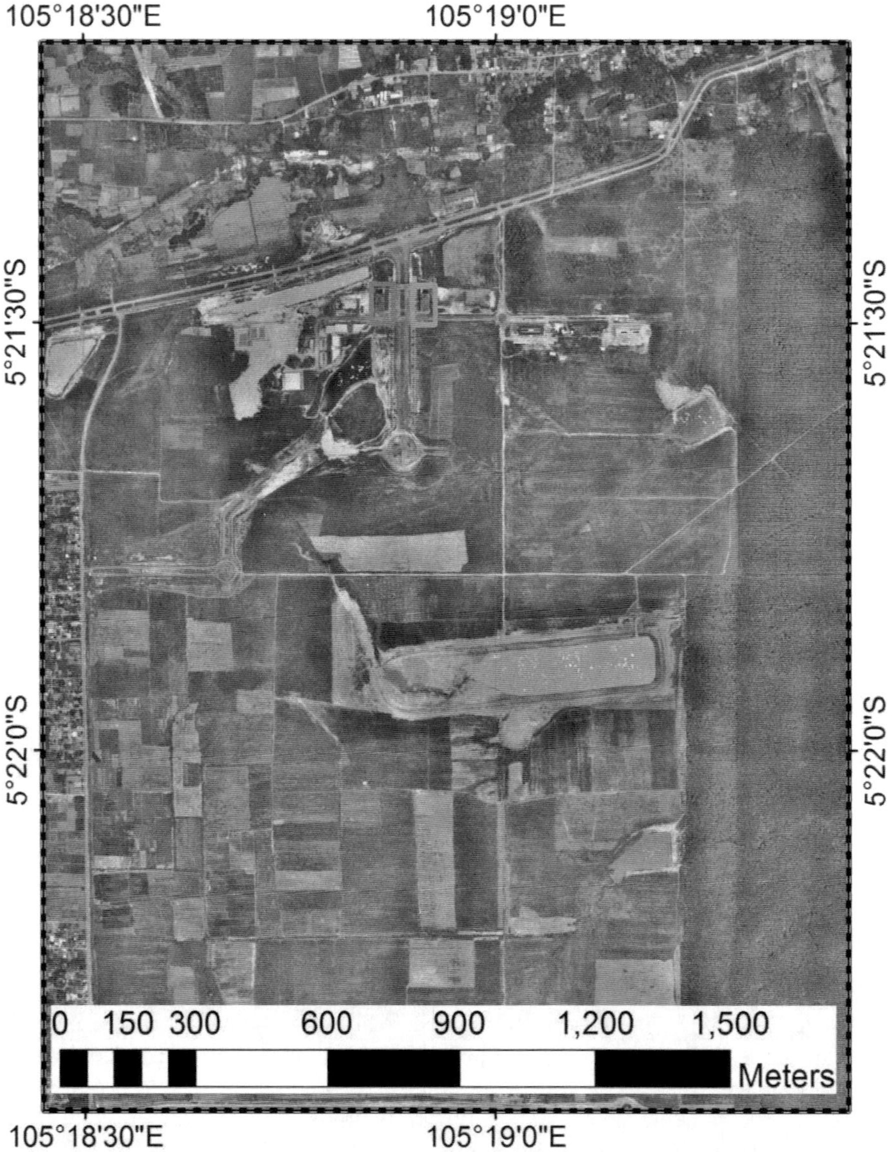

Figure 1. Location of case study. Gedung A is pointed by red rectangle while Gedung B is pointed by blue rectangle

2. Method

The process in giving value of facilities in room begins with collecting facilities data which related to attribute data and defining room boundary which related to spatial data. The process starts with data management and data processing. Tool used in applying multipurpose cadastre to evaluate energy security in this case energy consumption of land parcel is geographical information system (GIS). GIS The advent of computerised databases and GIS technology provides an opportunity to develop a greater understanding of land and how land may be more efficiently and effectively managed [4].

2.1. Data Management Method

Data collected in applying multipurpose cadastre are the fundamental component of multipurpose cadastre. Those are spatial reference frame, base maps, cadastre map, attributes data. Spatial reference frame is collected from 13 bench marks located in Institut Teknologi Sumatera referred to geospatial reference frame SRGI2013, reference ellipsoid WGS84 and reference frame ITRF2008 epoch 2012.0. Current base map is obtained from photogrammetry measurement with scale 1:1000 acquired on February 2017. Cadastre map that delineate boundary of Institut Teknologi Sumatra parcel is obtained from Institut Teknologi Sumatera land document. The principal boundary of this research is delineation of boundaries of Gedung A and Gedung B which is obtained from Institut Teknologi Sumatera building floor plan. Attributes data which is facility, occupant, and electronic devices are obtained by doing 'facility counting' activity.

Those attributes data are stored in database to facilitate updating, modifying, deleting the data. Those attributes data are classified into three entities: Facility, Room, and Facility-Room. Facility entity contains code, name, brand, figure, and category of facility. Category of facility is mark to distinguish whether the facility is electronic devices. Room entity contains code, name, usage, location, floor, and activity hour per day. Activity hour per day is mark to identify occupant activity in the room. Facility-Room entity contains facility code, room code which is constraint to code in facility entity and room entity, respectively, and number of good-conditioned facility, bad-conditioned facility, and unused facility in corresponding room. The code of facility and room are unique. Entity relationship diagram of this research is shown on **Figure 2**.

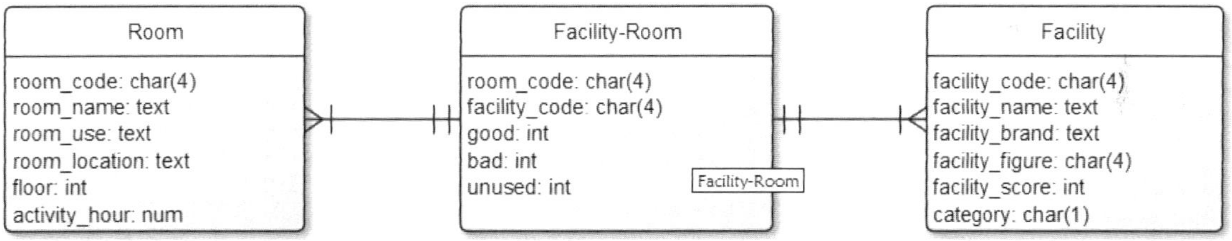

Figure 2. Entity relationship diagram of this research

The boundaries of Gedung A and Gedung B which is obtained from Institut Teknologi Sumatera building floor plan are rectified in Institut Teknologi Sumatera base maps with control points used are 4 corners of respective building. Area of room is calculated to provide weight in room valuation.

2.2. Data Processing Method

Exceed of energy consumption of every room are analysed by firstly calculate room value based on facilities beside electronic devices and electronic devices itself. Valuation and weighing for every facility is done depends on objective of valuation [5]. The objective of valuation in this research is to obtain energy consumption on respective room. It correlates with productivity of respective room. Less productive the room, it should be less energy consumed. Productivity of room can also be used to value a room [6]. Facility both electronic devices and general facilities valuation method is rating method based on its functionality.

The method of valuation is by comparing one facility to other facilities. General facilities included in room valuation are relatively static items (such as chair, desk, and cabinet) except lamps and other electronic devices. Electronic devices included in room valuation are air conditioner, CCTV, computer, dispenser, fan, lamps speaker, and other machines. The room value and number of electronic devices which consume electrical energy are correlated with each other. Consumption of electrical energy of electronic devices in the room with average value remains constant while consumption in other room needs to be evaluated to save the energy.

3. Discussion

The results of valuation are room values calculated both from general facilities and electronic devices. Differences in value for same room then are evaluated by comparing with other room value.

3.1. Room Value Analysis

Total room which is calculated in this research are 40 rooms in 2 buildings with 2 floors in respective building. Display of result is based on the location of room inside building and its floor resulting 4 displays: first floor of Gedung A, second floor of Gedung A, first floor of Gedung B, and second floor of Gedung B. Each display shows only either value calculated from general facilities or electronic devices. The value of room and its category is shown on **Table 1**. Displays of room value of first floor of Gedung A, second floor of Gedung A, first floor of Gedung B, and second floor of Gedung B are shown on **Figure 3, Figure 4, Figure 5**, and **Figure 6.**

Table 1. Room value based on general facilities and electronic devices

Room Code	Room Name	Value based on general facilities	Category based on general facilities	Value based on electronic devices	Category based on electronic devices
A101	Toilet Pria	13.0	Low	7.8	Low
A102	Toilet Wanita	0.0	Low	9.4	Low
A103	Ruang Panel	4.9	Low	4.1	Low
A104	Gudang	48.4	Low	124.4	Medium
A105	A104	391.7	High	324.2	High
A106	A103	156.0	Medium	278.9	High
A107	A102	160.0	Medium	343.0	High
A108	Server	18.2	Low	39.3	Low
A109	A101	334.9	High	649.8	High
A110	Bilik Toilet Wanita	41.1	Low	0.0	Low
A111	Bilik Toilet Wanita	37.8	Low	0.0	Low
A112	Bilik Toilet Pria	39.6	Low	0.0	Low
A113	Koridor Gedung A	65.2	Low	86.4	Medium
A201	A206	99.5	Medium	36.9	Low
A202	A207	16.0	Low	16.6	Low
A203	A208	242.7	Medium	455.4	High
A204	A205	470.3	High	733.6	High
A205	A204	138.0	Medium	84.8	Medium
A206	A201	157.6	Medium	174.7	Medium
A207	A202	197.7	Medium	331.1	High
A208	A201	209.9	Medium	142.6	Medium
A209	Koridor Gedung A	12.0	Low	124.6	Medium

IOP Conf. Series: Earth and Environmental Science **124** (2018) 012003 doi:10.1088/1755-1315/124/1/012003

Table 1 (cont.). Room value based on general facilities and electronic devices

B101	Ruang Panel	3.7	Low	3.2	Low
B102	Toilet Wanita	44.0	Low	9.4	Low
B103	Toilet Pria	41.2	Low	2.2	Low
B104	B101	98.5	Medium	385.5	High
B105	B102	389.1	High	247.8	Medium
B106	B103	7.0	Low	54.1	Low
B107	B104	123.9	Medium	314.5	High
B108	B105	141.1	Medium	338.5	High
B109	Bilik Toilet Pria	69.0	Low	0.0	Low
B110	Bilik Toilet Wanita	36.9	Low	0.0	Low
B111	Bilik Toilet Wanita	40.5	Low	0.0	Low
B112	Koridor Gedung B	74.3	Medium	49.4	Low
B201	B202	149.6	Medium	151.8	Medium
B202	B201	130.5	Medium	207.6	Medium
B203	B203	65.6	Low	196.2	Medium
B204	B204	88.1	Medium	211.6	Medium
B205	B205	112.6	Medium	261.5	High
B206	Koridor Gedung B	170.0	Medium	411.9	High

Figure 3. First floor of gedung A value calculated from general facilities (left) and electronic devices (right)

Figure 4. Second floor of gedung A calculated from general facilities (left) and electronic devices (right)

Figure 5. First floor of gedung B calculated from general facilities (left) and electronic devices (right)

Figure 6. Second floor of gedung B calculated from general facilities (left) and electronic devices (right)

The classification of room value falls into three categories: high, medium, low for either facility based value or electronic devices based value. Based on those figures and category, room with excess energy can be identified such as room with yellow (medium) category in facility which becomes green (high) category in electronic devices. Each room falls into one category because of cumulative reason. Reasons are number of general facilities, number of electronic devices, occupant, and area of respective of room. For example, room B106 has low facility based value because of lack of facility and low activity hour. Room A204 has high facility based value because of considerably facility and high activity hour. Room B206 has high electronic devices based value because there are many electronic devices that possibly to consume much energy. Room B102 which is toilet room has low electronic devices based value because there are few electronic devices mean less energy consumption.

3.2. Energy Consumption

Consumption of electrical energy of electronic devices in the room with average value remains constant while consumption in other room needs to be evaluated to save the energy. It is identified by looking at facility based value. In this research facility based average value is 116.0 and electronic devices based value is 170.3. Simply, it means room with facility based value below 116.0 should have electronic devices based value below 170.3 to save energy. Excess of energy is calculated by subtracting difference of each room electronic based value and average electronic based value with difference of each room facility value and average facility value.

The result of this research shows that room value correlate weakly with number of electronic device in corresponding room. It shows excess energy consumed in low value-room. Correlation between facility based value and electronic devices based value sorted by facility based value is shown on **Figure 7**. The category of excess energy in each room with excess energy is shown on **Table 2.** Room which is not shown on **Table 2** means energy used in corresponding room is secure.

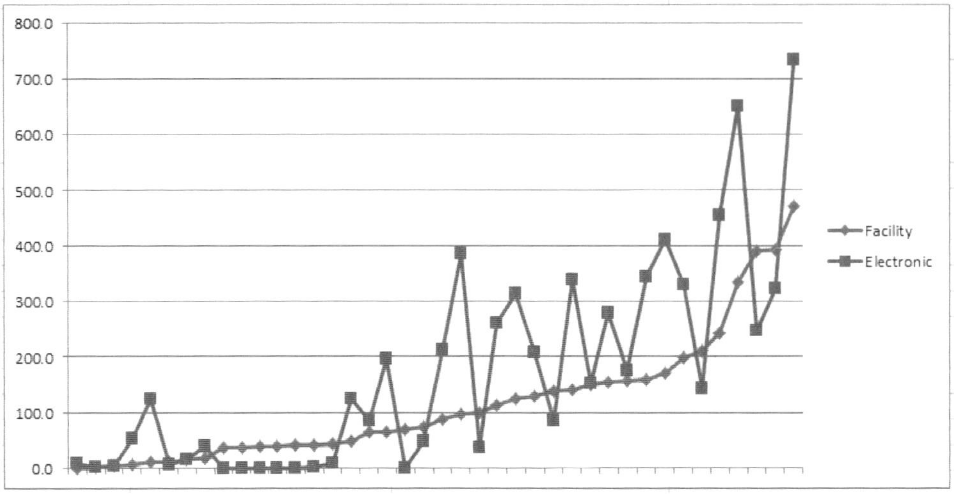

Figure 7. Correlation between facility based value and electronic devices based value

Table 2. Excess energy in room

Room Code	Value Difference	Category
A109	260.6	Critical
B104	232.7	Critical
A204	209.0	Critical
B206	187.6	High Need
A203	158.4	High Need
B108	143.1	High Need
B107	136.2	High Need
A107	128.7	High Need
B205	94.6	High Need
A207	79.1	Low Need
B203	76.2	Low Need
B204	69.2	Low Need
A106	68.6	Low Need
A209	58.3	Low Need
B202	22.8	Low Need
A104	21.7	Low Need

The classification of excess energy falls into three categories: critical, high need, and low need. Theoretically, there are some treatments to diminish excess energy of room. More critical the excess, more treatment needs to be done. Those treatments are:

1. Minimize energy consumption in the room. Although numbers of electronic devices do not always mean the consumption of electrical energy and there are plenty electronic devices, it is recommended for occupant to be careful in utilizing electronic devices in low-value room to minimize energy consumption.
2. Move electronic devices to other broader room. One of the factors in evaluating room value is room area. High electronic based value room often have many electronic devices with less facility because of not enough space in corresponding room.

3. Increase general facility in the room. If the room still have enough space to provide general facility, it will increase the productivity of corresponding room and make energy consumption more secure.

4. Conclusion

The method to evaluate excess energy consumption of room based on productivity and electronic devices produce classification of room with excess energy. The classification could lead to policy and strategies of energy security. The process started with concept of multipurpose cadastre: every land parcel has coordinates, boundary, and attribute data that show facility information and ends with evaluation of energy security. In this case study, there are 3 rooms with critical excess energy, 6 rooms with high excess energy, 7 rooms with low excess energy, and 24 rooms with secure energy. Although numbers of electronic devices do not always mean the consumption of electrical energy and there are plenty electronic devices, it is recommended for occupant to be careful in utilizing electronic devices in low-value room to minimize energy consumption.

Acknowledgment

This research was funded by Institut Teknologi Sumatera Smart Mandiri 2017 Research Grant Fund No. 235/IT9.KPA/SK/PP/2017. Authors acknowledge assistance from students of Institut Teknologi Sumatera to help collecting data for this research. Authors also acknowledge cooperation from every Institut Teknologi Sumatera staffs.

References

[1] Dale P F and McLaughlin J D 1988 *Land Information Management: an Introduction with Special reference to Cadastral Problems in Third World Countries* Oxford: Clarendon Press
[2] Pesl I 2003 Cadastre and Other Public Registers: Multipurpose Cadastre or Distributed Land Information System *FIG Working Week*
[3] Council N R 1983 *Procedures and Standards for a Multipurpose Cadastre* National Academy of Sciences Washington DC p 1
[4] Dale P F and McLaren R A 1999 GIS in land administration *Geographical information systems* **2** 859-875.
[5] Malczewski J 1999 *GIS and Multicriteria Decision Analysis* New York: John Wiley & Sons, Inc.
[6] Boyd T and Boyd S 2011 Valuing The Land Component Of Improved Investment Property For Tax Purposes *17th Pacific Rim Real Estate Society Conference Paper*

Green Building Premium Cost Analysis in Indonesia Using Work Breakdown Structure Method

V Basten[1,2]*, Y Latief[2], M A Berawi[2], Riswanto[2], and H Muliarto[1]

[1] Civil Engineering Department, Institut Teknologi Sumatera, Lampung Selatan, Lampung, 35365, Indonesia
[2] Civil Engineering Department, Faculty of Engineering, Universitas Indonesia, Kampus Baru UI, Depok, 16424, Indonesia

Email: van.basten@si.itera.ac.id

Abstract. The concept of green building in the construction industry is indispensable for mitigating environmental issues such as waste, pollution, and carbon emissions. There are some countries that have Green Building rating tools. Indonesia particularly is the country which has Greenship rating tools but the number of Green Building is relatively low. Development of building construction is depended on building investors or owner initiation, so this research is conducted to get the building aspects that have significant effect on the attractiveness using The Green Building Concept. The method in this research is work breakdown structure method that detailing the green building activities. The particular activities will be processed to get the cost elements for the green building achievement that it was targeted to improve better than conventional building. The final result of the study was a very significant work package on green building construction in the city of Indonesia case study.

1. Introduction

Economic growth development is supported by infrastructure development in some countries to meet the social and environment needs. Annual average amount of investment in infrastructure is up to US $4.6 trillion of the total construction in the world. Infrastructure construction contribute up to 8-10% total of Gross Domestic Products (GDP) in the world [1]. Furthermore, sustainable development has to be applied so that balancing the supply and demand in the countries could be achieved. Infrastructure development has the positive impact toward the economic aspect, but it has the negative impact to the environment aspect. In addition, increasing number of GDP will increase the number of high-risk building and Green House Gas (GHG) emission that it indicates the environment quality. Increasing the number of building activities in construction and operational, have impact on increasing number of world carbon emission. In 2007, carbon emissions in the world reach 29.65 billion ton or equivalent to 40% of total green house gas emission in the world. Then, the value is predicted increase up to 42.4 billion ton in 2035 [2, 3]. On the other hand, the condition is aggravated by exploitation of natural resources up to 30% to meet building construction and operational needs [4, 5]. At the end, all of the resources will produce the waste that it has recorded about 25-40% waste production sources in the world.

Indonesia as a particular development country in the world as the forth country with high number in building construction growth during 2009 to 2011 [6]. Furthermore, Indonesia have to control the development strategy to meet sustainability infrastructure. Indonesia is committed in world carbon emission reduction program, that is about 29% in 2030. There are some strategies in Indonesia to support the emission reduction progrom such as land use planning, energy and water efficiency program, effectiveness of natural resources, and preventive program in regional development towards climate change. In building development, Indonesia has Green Building Concept Program which it was a mandatory from the government policies. In fact, the number of green building in Indonesia is grown very slowly if it is compared to the other countries as described in Table 1. Indonesia is the lowest in growth number of green building toward the countries which have local green building rating tools. United state as the highest number of green building with LEED rating tools that was developed since 1998. After that, Hongkong with BEAM Plus rating tools reached 1,500 building certification since 2005 until 2014. In South East Asia, Singapore and Malaysia is still aboved Indonesia with more significant growth number than Indonesia which had Greenship as green building rating tools.

Table 1. Average Number of World Green Building Growth [3, 5-8]

Country Name	Year	Number of Building	Avg. Growth Number (per year)
US	1998-2012	13,000	867
Hongkong	2005-2014	1,500	150
Singapore	2005-2015	1,101	100
Germany	2009-2016	490	61
India	2001-2015	754	54
Malaysia	2009-2015	335	48
Japan	2001-2015	450	30
South Korea	2008-2012	53	11
Israel	2005-2013	45	5
Indonesia	2009-2015	23	3

Four main foundations in green building concept are mitigating the negative impacts of buildings activities on the environment, enhancing indoor health and comfort, payback period of the green features investation, and the effectiveness of green building concept in building life cycle. In Indonesia, based on the previous survey which the main reason of low interest in green building concept is higher cost of building features to achieve green concept [9]. Furthermore, there needs advance research identifying the cost component of green building. Work Breakdown Structure (WBS) approach is one of the systems that are needed in order to visually the work in order to complete a project. So, the final result of the breakdown is the work packet which could define the cost component [10]. This research is focus on defining the work packet and cost component of green features for green building concept development. Cost components which increased the total cost of building development could be reduced with the other alternatives in the planning phases without eliminates the function.

This research was carried out to solve two major problems, there are identifying the green building rating tools that affect to total building cost and evaluating the green building features need based on building certification classification. Therefore, this research observed new green building only because there was premium cost data record. In Indonesia, there are 12 new green building and 8 existing green building. The new building is valued until final assessment before operational and maintenance phase, beside that the existing building is valued with the retrofitting cost of conventional building toward green building. Observations were constrained by building function as office buildings in Jakarta with Greenship Certification.

2. Green Building

Introduction of green building concept was begun after the world oil crisis in 1970. Green building movement and revitalization was made over decades that focused on climate change issues, natural resources conservation, and urban development [11]. Furthermore, definition of green building is friendly environment concepts on building that design, build, and operate are conducted to minimize the negative impact on environment. Ensuring the green standard in some regions was on the track, some regions in the world made the green building rating tools. The first founded of green building rating tools is Building Research Establishment Environmental Assessment Method (BREEAM) in United Kingdom in 1990 [12]. Rating tools are not only provided to meet the environment needs but also the cost effectively in adopting sustainable solutions [13]. There are several aspects that are assessed in green building certification process in several rating tools in the world such as site development, transport, energy conservation, water efficiency, material resources, indoor health and comfort, building management, pollution, waste, innovations, regional priority, and economic.
In Indonesia, there are Greenship rating tools as the local standard for green building certification. Greenship has six aspects in building assessment such as Appropriate Site Development (ASD), Energy Efficiency & Conservation, Water Conservation, Material Resources & Cycle, Indoor Health & Comfort, and Building & Environment Management [3]. Table 2 describes the aspects compilation on green building rating tools in the world. There are twelve aspects that were decided in some countries assessment according to the regional needs. Indonesia adopts six aspects on green building rating tools that are named Greenship. Greenship is under coordination of World Green Building Council, so that the point of assessment is not too different to LEED rating tools. Based on the average weighting score of the world green building rating tools aspect, Energy Conservation has the highest score. After that, the second position of the highest score achievement of green building rating tools are Indoor health & Comfort and Site Development.

Table 2. World Green Building Rating Tools & Weighting Score [3, 6, 14, 15]

Rating Tools	BREEAM 2013	LEED 2009	CASBEE	HK-BEAM 4/04	Green Star	Green Mark 4.1	IS Standard 5281	Green Building Index 2009	DGNB 2011	Greenship Ver. 1.2	Average
Found Year	1990	1993	2001	2002	2003	2005	2005	2009	2009	2009	
Country	UK	USA	Japan	Hongkong	Australia	Singapore	Israel	Malaysia	Germany	Indonesia	
Site Development	4.5%	23.6%	16.7%	13.9%	7.0%	22.1%	9.8%		22.5%	16.8%	14.4%
Transport	8.1%				9.0%		1.7%				1.7%
Energy Conservation	27.0%	31.8%	16.7%	41.7%	25.0%	61.1%	32.1%	35.0%		25.7%	28.6%
Water Efficiency	8.1%	9.1%	8.3%	6.7%	14.0%	8.9%	15.7%	10.0%		20.8%	10.6%
Material Resources	10.8%	12.7%	8.3%	12.8%	13.0%		7.5%	11.0%		13.9%	9.6%
Indoor Health & Comfort	9.0%	13.6%	16.7%	25.0%	15.0%	4.2%	15.3%	21.0%	22.5%	9.9%	14.5%
Building Management	19.8%		16.7%		9.0%		7.8%	16.0%	10.0%	12.9%	9.4%
Pollution	4.5%		16.7%		8.0%						3.3%
Waste	6.3%						1.4%				1.1%
Innovation/ Green Features	1.8%	5.5%				3.7%	8.6%	7.0%	22.5%		4.5%
Regional Priority		3.6%									0.3%
Economic									22.5%		2.0%
Max Point	100%	100%	100%	100%	100%	100%	100%	100%	100%	100%	100.0%

3. Premium Cost

Several benefits using green building concept are not only in environment aspect but also in social and economic aspects. Using sustainable concept in building enhances building performance especially in building life cycle such as procurement, installation, operational, waste disposal, and recycle process

[16]. In fact, starting the high-performance property development is not easy because it requires relatively higher ability especially in financing and knowledge than conventional building. Additional cost is required for green building concept implementation named premium cost. Premium cost is a cost used to implementation green features so that building efficiency could be reached in operational and maintenance phase [17].

In California, green building implementation required premium cost up to 1.84% or more expensive US $43 per sqm than conventional building cost. The average premium costs data are collected from 33 buildings in California as an office building function [18]. On the other hand, green building in Australia required up to 3-5% premium cost for a 5-star new green building development. Based on the country standard in Australia, minimum building achievement is equal to 4 stars [19]. In addition, the green building premium cost to meet a BREEAM Excellent rating was up to 6% for green features. The previous research about green building explained that green building concept has a higher initial cost, longer in design, and using non-standard materials or system. Beside that, green building concept has a benefit in long-term cost such as energy, water, and resources savings in operational and maintenance phase.

4. Work Breakdown Structure

Work Breakdown Structure (WBS) approach is a visual form of work that could be used to describe the project scope [10]. The main purpose of WBS is to effectively decompose the project requirement from planning until monitoring and controlling phase. In addition, WBS could be the material of lesson learned for the next project as historical information. WBS helps in detailing the project deliverable and activities that are broken down as manageable parts in resources and costs estimation, scheduling, and controlling each phase of a project. WBS is effective in bigger project or involves the complicated elements of construct ruction. Decomposition process in WBS is a process of subdividing a project requirement into smaller component by all of the project team. Comprehensive WBS could be reached if the team work has good experience, communication, and high integrity.

WBS process is started with define the project scope. Project scope is broken down into particular activity that is named work packages. Graphical illustration of WBS helps the project stakeholder to divide the project scope based on management component levels and deliverables. After that, the deliverables in a bigger project is divided by planning package of progressive elaboration. The entire work package is under a control account for monitoring and controlling the process is in line with the project objectives and achievements. Work package is the smallest level of deliverables in WBS, but the number of work package depends on the characteristics of a project. Otherwise, the project manager needs is used to get the accurate estimating and effective project monitoring [20].

5. Research Methodology

This research was conducted in three stages. At the first stage, an identification aspects of green building rating tools that contributed to building premium cost, was conducted by qualitative approach used in depth interview. There are eight aspects of green building rating tools that was tested such as Appropriate Site Development, Energy Efficiency & Conservation, Water Conservation, Material Resources & Cycle, Indoor Health & Comfort, Building Environment & Management, Innovation, and Building Life Cycle. Based on the literature, 272 indicators of green building aspects or independent variables have a potential impact on green building premium costs. Beside that, there are 40 indicators of dependent variables that building cost component was changed because of green building concept. Five people is selected as an expert in this research that had greenship professional certificate issued from Green Building Council Indonesia. Three experts work in consultant firm and two others work in Developer Company with more 10 years of experience. 118 of 272 independent variable indicators were chosen by the experts.

The second stage is evaluation of the variables that affected to green building premium cost using questioner as an instrument. The questioner is distributed to 47 respondents with related stakeholder with a green building development background, including developer and consultant bodies with 10 years of experience. The instrument used closed questionnaire with Guttman Scaling, so the respondents have just chosen "Yes/ No." For instrument validation, Coefficient of Reproducibility is calculated from

$$C_r = 1 - \frac{e}{n} \tag{1}$$

where e = total errors and n = the number of responses. The number of responses (n) are equal to the number of respondents (N) multiplied by the number of variables (k). The coefficient of reproducibility measures how well the respondents responses from his position within the statement. C_r should be at least 0.90 [21]. After that, completed the validation process, Coefficient of Scalability is given by

$$C_s = 1 - \frac{e}{c(n-T_n)} \tag{2}$$

where e = total errors, c = probability the true answer or it equal to 0.5; n = the number of responses, and T_n= the number of choices [22]. The level of acceptance Cs may be 0.60. If the amount of Cr is less than 0.90 or Cs is less than 0.60, the most influential variable on decreasing the total errors is deleted from the analysis. Beside that, the characteristics of respondent can be used to improve the amount of Cr and Cs. It is reasonable to eliminate the highest error on person's data because the different characteristics of respondent can make low accuracy data [23].
The next test on questionnaire instrument is reliability measurements. Kuder-Richardson Formula 20 or KR-20 is a measurement reliability to check the consistency results of binary variables. KR-20 is used because the indicators of green building rating tools have varying difficulty for implementation. This is the formula of KR-20

$$ri = \frac{k}{(k-1)} \left(\frac{S_t^2 - \Sigma pi\, qi}{S_t^2}\right) \tag{3}$$

where k = sample size for the test, S_t^2= variance of the whole test (standar deviation square), Σpi qi = sum the product of pq for all n items, p = proportion of people passing the item, and q = proportion of people failing the item or 1-p. true item Range scores of reliability are from -1 to 1, where -1 is low reliability and 1 is perfect reliability [24].
Data analysis from questioner is simplified to get the conclusion of green building rating tools that affected to green building premium cost. This research used descriptive statistics to describe the basic features of the data. The descriptive statistic that used for nominal data is the frequency distribution with proportion percentages[25]. To calculate the percentage of frequency distribution could be defined as

$$P = \frac{f}{n} \times 100\% \tag{4}$$

Where f = frequency of the chosen answer, n = the number of responses. After that, the percentage of the frequency data distribution is interpreted into the category such as 0 – 1% = no impact, 2% - 25% = very little impact, 26% - 49% = little impact, 50% = middle impact, 51% - 75% = big impact, 76% - 99% = very big impact, and 100% = totally impact [26]. The green building aspects which have score greater than 50%, are chosen as the indicator of the green building premium cost components.
In the last stage, this research conducted visualization of green building scope that has impact to the premium costs used WBS. The scope of project is broken down based on the data which were collected using in depth interview to the building owner or building management corporate. The WBS describes the cost components percentage and the work packages in the initial building development process. The sample of building data to develop the WBS consists of 2 buildings with platinum

certification, 2 buildings with gold certification, and a building with silver certification. Total 5 of 12 new green building that was certified by Green Building Council Indonesia, was analysed used average percentage premium cost in WBS visualization.

6. Result and Discussion

There are eight independent variables and one dependent variable in this research. Independent variables are consisted of appropriate site development (X1), energy efficiency and conservation (X2), water conservation (X3), material resources and cycle (X4), indoor health and comfort (X5), building environment and management (X6), innovation (X7), and building life cycle (X8). On the other hand, increasing factor of the green building rating (Y) is the dependent variable in this research. Based on the instrument validation, 40 of 118 independent sub variables and 13 of 40 dependent sub variables have been chosen as the critical factors of rating tools that affects the green building premium cost. Table 3 describes the process of Coefficient of Reproducibility (Cr) and Coefficient of Scalability (Cs) Calculation using Eq. (1) and Eq. (2). All of the Cr scores are bigger than 0.90 and Cs scores are bigger than 0.60, so the questionnaire instrument is valid.

Table 3. Process of Cr and Cs Calculation for Instrument Validation

Variables	X1	X2	X3	X4	X5	X6	X7	X8	Y
e	90	110	36	32	88	84	20	22	174
N	47	47	47	47	47	47	47	47	47
k	20	24	9	9	21	23	6	9	40
n	940	1128	423	423	987	1081	282	423	1880
Cr	0.9043	0.9025	0.9149	0.9243	0.9108	0.9223	0.9291	0.9480	0.9074
Tn	190	249	134	97	220	191	49	90	331
Cs	0.76	0.75	0.75	0.80	0.77	0.81	0.83	0.87	0.78
Subvar.	6	8	4	4	7	6	2	3	13

The next step is to calculate the reliability score using the formula of KR-20 on Eq. (3). The reliability test is conducted to 40 independent sub variables and 13 dependent sub variables. Total of 53 variables are the sample size validated in advanced. Sum the product of pq are accumulation of the proportion between the people who agree and disagree which the sub variable impacted to the green building premium cost. The result is described on Table 4 in which the instrument has high reliability with score 0.85 (between 0.71 and 0.90). After all of the validation and reliability test of the research instrument are eligible, this research continued to the frequency distribution calculation and description based on the categorization of the variables and they are followed with the evaluation using WBS.

Table 4. Process of KR-20 for Instrument Reliability Test

Element	Value
K	53
$\Sigma p_i q_i$	11.71824

IOP Conf. Series: Earth and Environmental Science **124** (2018) 012004 doi:10.1088/1755-1315/124/1/012004

Table 4. (Continued)

Element	Value
s_t^2	73.89316
Mean	32.97872
KR-20	0.85760

6.1. Appropriate Site Development (ASD)

Appropriate Site Development is the first variable that is evaluated in Greenship rating tools. Based on the literature study, there are 20 sub variables in ASD. Six of twenty sub variables in ASD have the impact to green building premium costs. Beside that, all of the chosen sub variables have the high category in enhancing building cost because more than 28 of 47 respondents are agree with the sub variables. There are landscape area (37), building envelope (34), landscape design (32), storm water technology (31), basic green area (28), and storm water management (28) which Figure 1 presents the percentage value of ASD sub variable. The average amount of respondents choosing on sub variable is 32 which there are three sub variables with the highest impact on green building premium cost. The first is landscape area which the premium costs were used for roof garden, terrace garden, and wall garden. The second is building envelope which the premium costs were used for green roof and avoiding material with heat island effect. The third is landscape design which it is consisted of kind of the softscape and hardscape.

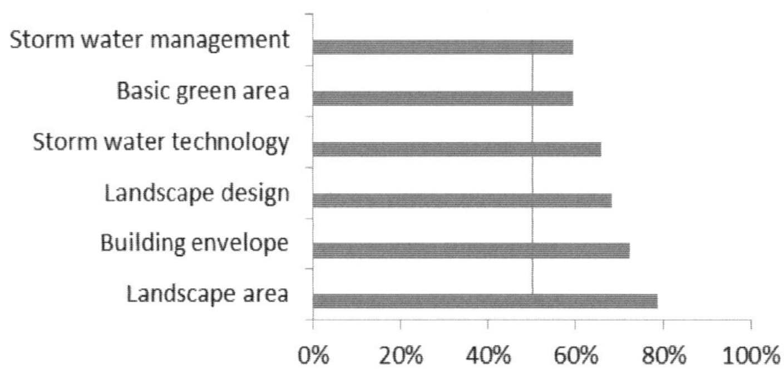

Figure 1. Sub variables of appropriate site development

6.2. Energy Efficiency and Conservation (EEC)

The second aspect or variable of green building is energy efficiency and conservation. Based on the respondent's opinion, there are 8 of 24 EEC sub variables that have an impact to enhancing the green building premium cost. All of the chosen sub variables have the big impact category such as energy savings lamp (37), automatic/ sensor system (36), ventilation system (36), lux sensor system (30), building envelope (28), CO_2 sensor (28), renewable energy sources (27), and thermal transmission (27) which Figure 2 describe the percentage value impact of EEC sub variable to green building premium cost. Three of eight sub variables have values above the average number of 32. The sub variable with the highest amount is energy saving lamps. Premium costs in this sub variable are used for lighting system which they are natural lighting system and artificial lighting. After that, the automatic sensor system on lamp is the second sub variable in EEC which above the average amount. The last one is ventilation system which the price is based on the technology used.

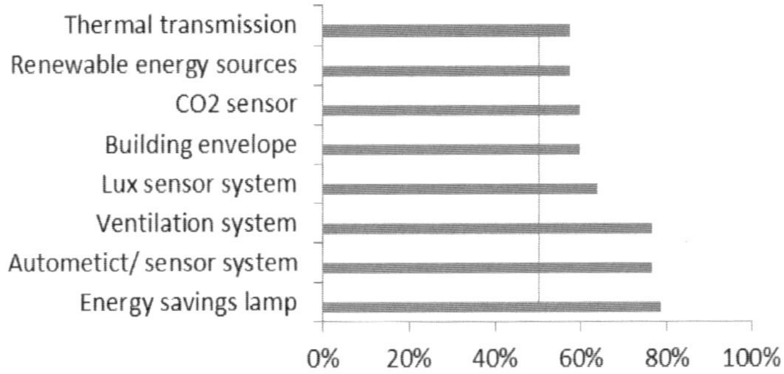

Figure 2. Sub variables of energy efficiency and conservation

6.3. Water Conservation (WAC)

Water conservation is consisted of 9 sub variables. Based on the survey result, there are four of nine sub variables that are impacted to the green building premium cost. All of the chosen sub variables have greater than 50% weighting score such as water treatment for AC (41), recycling water (34), water controller in landscape area (31), and water treatment for alternative sources (28). Figure 3 shows the rank of achievement of WAC variable. The average value of water conservation sub variable is 34. There are two sub variables equal or above this average amount. The first is recycling water system which the premium cost is used for recycling water machines implementation. The next is water treatment for air conditioning which it is more advance than the recycling system because the water is recycled for air cooler or chiller.

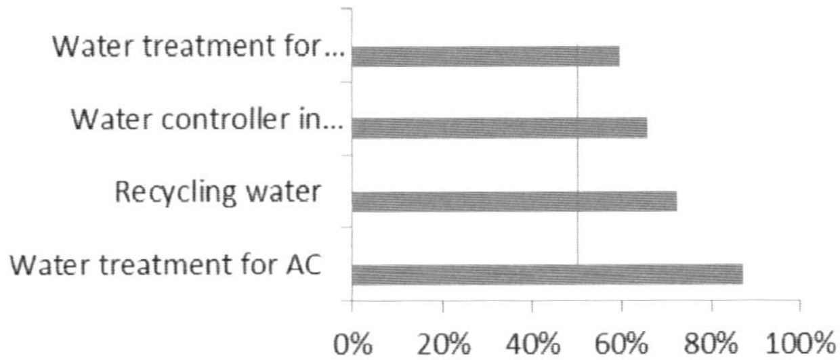

Figure 3. Sub variables of water conservation

6.4. Material Resources and Cycle (MRC)

There are nine of basic sub variables of material resources and cycle criteria. Based on the questioner result, there are four of nine that have direct impact on green building premium cost. Only one variable has high impact on MRC aspect, there is certified friendly environment material. Figure 4 describe the frequencies of MRC sub variables. Beside that, there is only certified friendly environment material sub variable which the value is above the average of respondent. The average respondent value of this MRC variable is 25. Using certified materials in Indonesia is more expensive than non-certified. There are an assessment value of the raw material resources, process, and legality. For example, using the certified wood based on forest stewardship council which it is comen from legal logging, green process, and local material.

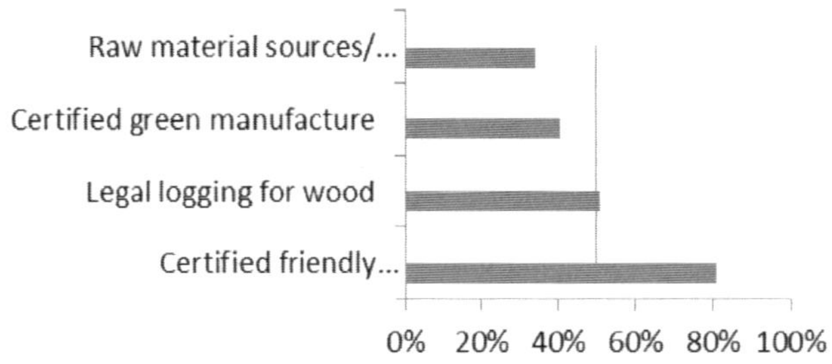

Figure 4. Sub variables of material resources and cycle

6.5. Indoor Health and Comfort (IHC)

Following the identification of indoor health and comfort sub variable, 7 of 21 sub variables was chosen by calculating the percentage of frequency distribution. The seven sub variables are insulation on building machine equipment (37), low VOC for wall material (34), humidity control (33), CO_2 control (32), indoor temperature (31), noise control (27), and adhesive control (26). Using insulation in some building equipment and machines are the highest impact to green building premium cost. The others sub variables have the highest impact to green building premium cost based on the respondents opinion using percentage of frequency distribution that is greater than 50%. Figure 5 describes all of the sub variables weighting score results for indoor health and comfort. The average value of respondent in indoor health and comfort sub variable is 32. There are insulation on building machine equipment, low VOC for wall material, humidity control system, and CO_2 control. The building developer spends the premium cost on the health and comfort control system.

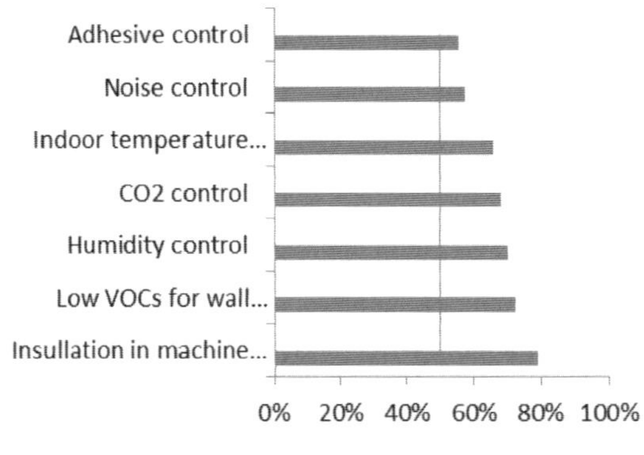

Figure 5. Sub variables of indoor health and comfort

6.6. Building Environment and Management (BEM)

Based on the literature review, for building environment and management aspects, there are 23 sub variables that calculated in green building assessment. The respondents in this research chose 6 of 23 sub variables that have impact to green building premium cost. All the six sub variables have big impact because the weighting score is greater than 50%. There are testing-commissioning periodically (37), dumpster (34), water flood controller (34), periodic testing of building installation (32), integration of building (28), and improvement action (26). Figure 6 describes the weighting scores of building environment and management sub variables. The category average amount of building environment and management is 32 consisted of testing-commissioning, dumpster, water flood controller, and periodic testing of building installation. In building environment and management, premium cost incurred to make building easy to maintain and to operate. Beside that it has to effective and efficient in building life cycle and friendly to the environment.

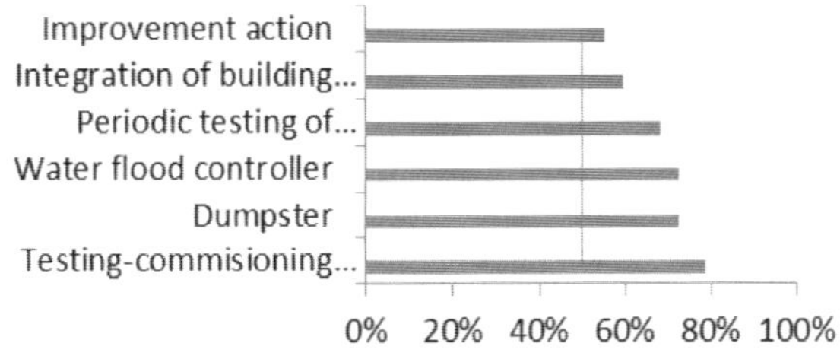

Figure 6. Sub variables of building environment and management

6.7. Innovation (In)

Innovation is one of the optional variables, because it is chosen based on literature study from other green building rating tools. There is not an innovation aspect in greenship, so for this study using innovation is the alternatives of variables that have potential impact to green building premium cost. There are two of six the biggest impact of innovation on green building premium cost such as environment improvement and cleaning self facade system. But, the biggest sub variables impact in innovation aspect toward green building premium cost is only environment improvement (Figure 7). The average value of innovation sub variable is 25. So, only one sub variable is above the average amount that is environment improvement. This sub variable is used for sustainability implementation to the building environment such as the internet of things, smart building, and others information technology use.

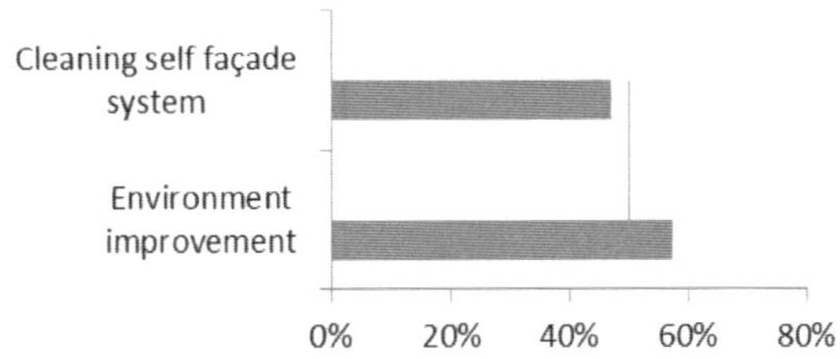

Figure 7. Sub variables of Innovation

6.8. Building Life Cycle (BLC)

The second alternative variable in this research is building life cycle aspect. According to the literature study, there are 9 sub variables in building life cycle aspect. The respondents chose 3 of 9 indicators that have impact in green building premium cost. There are technology update, building operational, preventive, and maintenance cost, and building access. All of the sub variables have big impact to green building premium cost because the weighting scores are greater than 50%. Figure 8 describes the all sub variables achievement on building life cycle variable. The average value of respondent for all sub variables is 30. In green building concept especially building life cycle, the premium costs are affected by technology update to redesign or retrofit the building components.

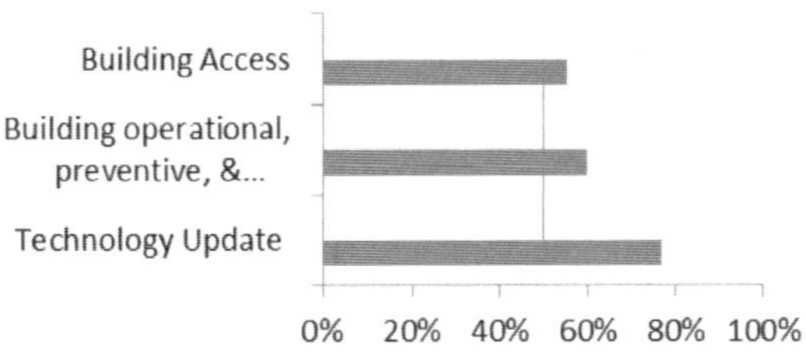

Figure 8. Sub variables of building life cycle

6.9. Enhancing Green Achievement by Premium Cost

Currently, some developers conducted improvement the building quality to meet green buildings concept because the government has the mandatory policies in the building development in Indonesia country. So that, these dependent variables were tested for green building cost evaluation that affected to enhance the green building achievement in Indonesia. There are 40 sub variables that have potential in green achievement. There 13 of 40 sub variables can be enhanced to improve the green building achievement (Figure 9). But, only 7 of 13 sub variables have the big impact such as energy savings (36), storm water management (33), lightening (29), alternative water sources (27), building facade (25), bicycle facility (24), and water reduction (24). The average amount of respondent in this variable is 26. The sub variables which are above the average value of respondent are consisted of alternative water resources, lighting, storm water management, and energy savings. This four sub variables are the higher potential of building assessment improvement by premium cost than others.

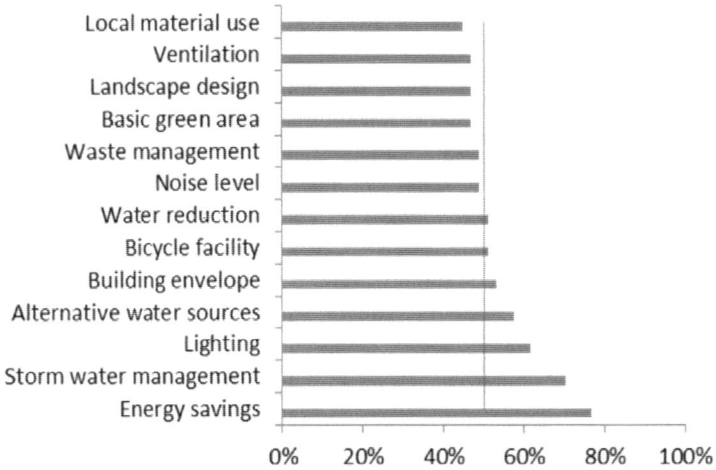

Figure 9. Sub variables of Enhancing Green Achievement

7. WBS for Premium Cost Evaluation

The last stage of this research is the work breakdown structure evaluation, The following scenario for developing green building features for achieve the rating tools is divided in 8 aspects. The scope of green building premium work in green building development is based on the previously identified with questioner and in depth interview. Based on the sub variables on the previous description makes the WBS like on Figure 10.

In Figure 10, P means green building certified platinum, G means Gold rating achievement, and S means Silver rating achievement. The written value on table is the weighting score of case study premium cost in green building Indonesia. DKI Jakarta as the capital city of Indonesia has the highest number of new green building in Indonesia which the amounts of building certified by green building council Indonesia are six buildings. But, this paper only contains 5 green building premium cost data because of the access limitation. This research used 2 platinum building, 2 gold building, and a silver building as the building case study. The highest green building premium cost in platinum building is the material resources and cycle aspect (48.13%); on the other hand the highest premium cost on gold certification building is the energy efficiency and conservation (92.76%). Then, the silver building used premium cost to meet building environment and management aspect.

The highest focus of green building developer is energy savings, friendly environment material, and building environment and material. They spend their money for energy savings using energy saving lamps, intelligent lighting control, and building envelope design and simulation. In material resources and cycle, the fullest attention in application is prefabrication material and using local raw material in construction. And the last, the building environment and management aspect pays attention on garbage technology and the composter process.

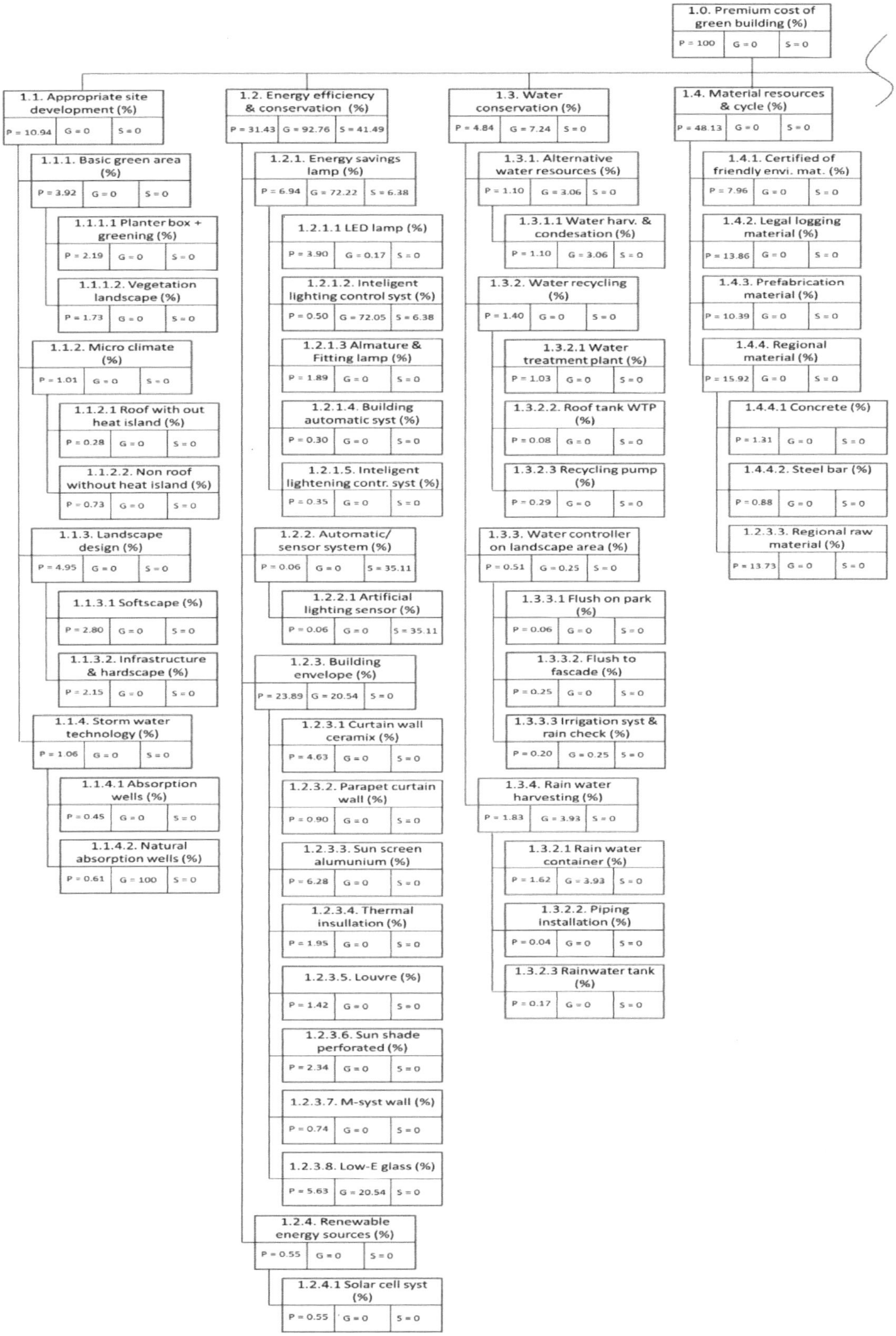

Figure 10. Work Breakdown Structure Based on Green Building Premium Cost

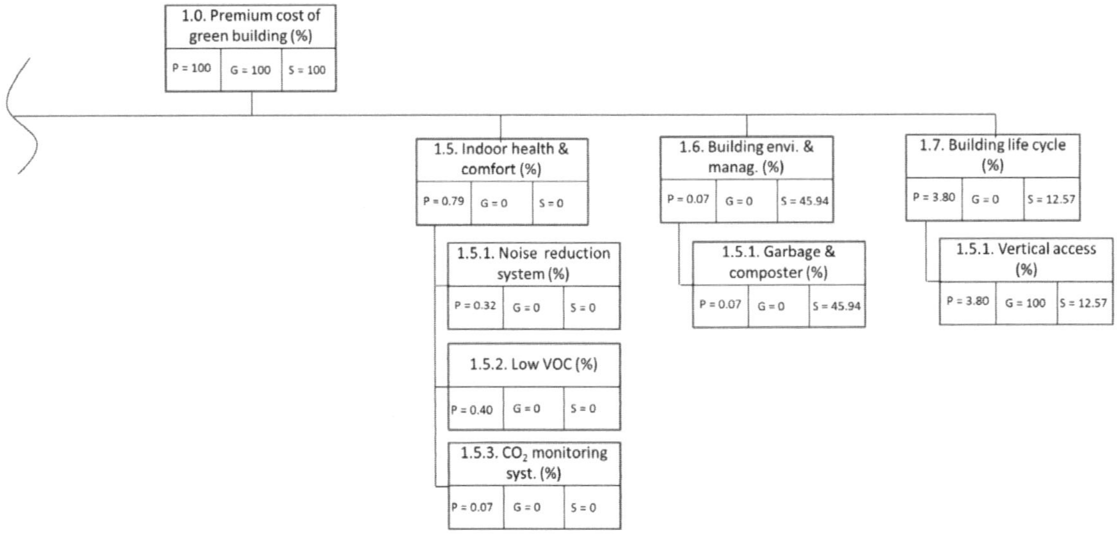

Figure 10. (Continued)

8. Conclusion

According to literature study and expert validation, there are eight independent variables with 118 of 272 sub variables that affected on green building premium cost. On the other hand, there are 40 of 118 sub variables of dependent variable. This research was continued with identification the most impact factors of applying green rating tools based on respondents that they involved in green building development. The respondents chose 53 of 118 independent sub variables that have high impact on green building premium cost. The top six of critical factors affected green building premium cost are water treatment for AC (or indoor usage), certified material of friendly environment, periodically testing-commissioning, insulation usage in building machine room, energy saving lamps, and basic green area.

Evaluating the green building features needs were depend on building certification classification. Almost all of green building features are applied in platinum green building. The highest weighting score of premium cost to achieve platinum certified green building in Indonesia are building envelope (23.89%), regional materials (15.92%), and legal logging material (13.86%). On the other hand, gold certified green buildings have different achievement that they tend to pay attention on energy in the green building premium cost (21.435) with sub–indicators of intelligent lighting control system features (72.05%) and building envelope (20.54%). Different things happen in silver certified green building in Indonesia which the only one building, so case study data also still limited. However, based on the sample data obtained that the silver green building in Indonesia concerned on building environment and management (45.94%) with garbage and composter technology. In addition, the other application of green features is in energy efficiency and conservation (41.49%) with automatic or sensor system (35.11%).

This research weighting scores are based on case study of green building in Indonesia. The improvement of green building development especially in cost and green features analysis could be based on this case study. The mandatory policy from government of Indonesia was already presented as one of the requirement to get the building permit. Unfortunately, this concept doesn't attractive because the premium cost is needed in the beginning of building development. Based on this paper result, there are some potential way to reduce the initial cost of green building such as defining the incentives rule which it was stated in government policy but not in details, modeling life cycle cost to get the effective of investment and incentives cycle, and greater technology competition are expected to breakthrough the enhancing of green building achievement especially to full fill the target in 2030.

Acknowledgement
The authors wish to thank for funders of this research that is ITERA MANDIRI Grants in Institut Teknologi Sumatera (Hibah Mandiri ITERA). Beside that, we thank to all the organizations and individuals who kindly participated in the research. Their time and input in the interview and survey are much appreciated. The authors would also like to thank and acknowledge Project Management Study Program, Faculty of Engineering, Universitas Indonesia, Indonesia, for generous support and encouragement of this research especially for the networking to green building colleague.

References

[1] M. Samari, N. Ghodrati, R. Esmaeilifar, P. Olfat, M.W.M. Shafiei, The investigation of the barriers in developing green building in Malaysia, Modern Applied Science, 7 (2013) 1.
[2] A.E. Outlook, Energy information administration, Department of Energy, 92010 (2010).
[3] Y. Latief, M.A. Berawi, V. Basten, R. Budiman, Riswanto, Premium cost optimization of operational and maintenance of green building in Indonesia using life cycle assessment method, AIP Conference Proceedings, AIP Publishing, 2017, pp. 020007.
[4] J.D. Linton, R. Klassen, V. Jayaraman, Sustainable supply chains: An introduction, Journal of Operations Management, 25 (2007) 1075-1082.
[5] M. Wimala, E. Akmalah, M.R. Sururi, Breaking through the Barriers to Green Building Movement in Indonesia: Insights from Building Occupants, Energy Procedia, 100 (2016) 469-474.
[6] V. Basten, Y. Latief, M.A. Berawi, R. Budiman, Riswanto, Evaluation of green building rating tools based on existing green building achievement in Indonesia using Life Cycle Assessment Method, AIP Conference Proceedings, 1818 (2017) 020008.
[7] Y. Wang, X. Wang, J. Wang, P. Yung, G. Jun, Engagement of Facilities Management in Design Stage through BIM: Framework and a Case Study, Advances in Civil Engineering, 2013 (2013) 8.
[8] H. Tan, S. Chen, Q. Shi, L. Wang, Development of green campus in China, Journal of Cleaner Production, 64 (2014) 646-653.
[9] H.-T. Nguyen, M. Gray, M. Skitmore, Comparative study on green building supportive policies of Pacific-Rim countries most vulnerable to climate change, DOI (2016).
[10] C.S. Snyder, A Guide to the Project Management Body of Knowledge: PMBOK (®) Guide, Project Management Institute, 2014.
[11] J.R. Wolch, J. Byrne, J.P. Newell, Urban green space, public health, and environmental justice: The challenge of making cities 'just green enough', Landscape and Urban Planning, 125 (2014) 234-244.
[12] K. Gowri, Green building rating systems: An overview, ASHRAE journal, 46 (2004) 56.
[13] J. Lowe, N. Watts, An evaluation of a Breeam case study project, Sheffield Hallam University Built Environment Research Transactions, 3 (2011) 42-53.
[14] Y. Wang, X. Wang, J. Wang, P. Yung, G. Jun, Engagement of facilities management in design stage through BIM: framework and a case study, Advances in Civil Engineering, 2013 (2013).
[15] M.A.B. Yusuf Latief, Van Basten*, Riswanto, Rachmat Budiman, Construction Performance Optimization toward Green Building Premium Cost Based on Greenship Rating Tools Assessment with Value Engineering Method, Journal of Physics, 877 (2017).
[16] L.N. Dwaikat, K.N. Ali, Green buildings cost premium: A review of empirical evidence, Energy and Buildings, 110 (2016) 396-403.
[17] C.J. Kibert, Sustainable construction: green building design and delivery, John Wiley & Sons2016.
[18] Y. Liu, X. Guo, F. Hu, Cost-benefit analysis on green building energy efficiency technology application: A case in China, Energy and Buildings, 82 (2014) 37-46.
[19] P. Morris, D. Langdon, What does green really cost, PREA quarterly, DOI (2007) 55-60.

[20] Y.H. Kwak, C.W. Ibbs, Project management process maturity (PM) 2 model, Journal of management in engineering, 18 (2002) 150-155.
[21] J.M. Linacre, Guttman coefficients and Rasch data, Rasch Measurement Transactions, 14 (2000) 746-747.
[22] H. Menzel, A new coefficient for scalogram analysis, Public Opinion Quarterly, 17 (1953) 268-280.
[23] J.R. Hills, F. King, Construct Validity of the Florida Functional Literacy Test, DOI (1978).
[24] G.F. Kuder, M.W. Richardson, The theory of the estimation of test reliability, Psychometrika, 2 (1937) 151-160.
[25] H. Oja, Descriptive statistics for multivariate distributions, Statistics & Probability Letters, 1 (1983) 327-332.
[26] W.M. Trochim, Qualitative measures, Research Measures Knowledge Base, 361 (2006).

Life Cycle Analysis for the Feasibility of Photovoltaic System Application in Indonesia

H M Yudha[1], T Dewi[2], P Risma[2], and Y Oktarina[2]

[1]Universitas Tridinanti Palembang, Palembang Indonesia,
[2]Politeknik Negeri Sriwijaya, Jalan Srijaya Negara Bukit Besar Palembang, 30139

E-mail: tresna_dewi@polsri.ac.id

Abstract. Electricity has become the basic need for everyone, from industry to domestic. Today electricity source still depends heavily on fossil fuels that soon will be diminished from the earth in around 50 years. This condition demands us to find the renewable energy to support our everyday life. One of the famous renewable energy sources is from solar, harnessed by energy conversion device named solar cells. Countries like Indonesia are gifted with an abundance of sunlight all the yearlong. The application of solar cells with its photovoltaic (PV) technology harnesses the sunlight and converts it into electricity. Although this technology is emerging very fast, it still has some limitation due to the current PV technology, economic feasibility, and its environmental impacts. Life cycle assessment is the method to analyze and evaluate the sustainability of PV system and its environmental impact. This paper presents literature study of PV system from the cradle to grave, it begins with the material choices (from the first generation and the possibility of the fourth generation), manufacturing process, implementation, and ends it with the after-life effect of PV modules. The result of this study will be the insights look of the PV system application in Indonesia, from the best option of material choice, the best method of application, the energy payback time, and finally the possible after life recycle of PV materials.

Keywords: Life cycle analysis, photovoltaics, renewable energy

1. Introduction

Modern life creates the improvement of life quality and has made electricity as the basic need for everyone although not everyone in a developing country or third world country can have electricity connection in their homes. Today's electricity depends heavily on fossil fuels that contribute a lot to greenhouse gas emission, therefore promotes more on global warming. Another fact is that the source of fossil cell will diminish in around 50 years [1]. This unfortunate condition has driven scientists to search for new and renewable sources of power. Renewable energy is a type of energy generated from a continuous and recurring source in nature/environment. The generation of renewable energy can be at the same rate of their usage. The renewable energy can come from the air, water, earth, and sun. One of this renewable energy is provided by the unlimited source of sunlight from the sun.

The utilization of sun-ray for heating had been done by the Romans in around first to fourth century [2]. Edmond Becquerel in 1839 found the effect of photovoltaic (PV), however 1954 was considered as the time when the photovoltaic technology was born, since on that day, Daryl Chapin, Gerald Pearson, and Calvin Fuller created a photovoltaic cell made of silicon and recorded as the first solar cell having ability to convert sun ray into electricity that could power an electrical equipment in Bell Labs, and since this innovation, PV system application had been increased tremendously [2][3]. Every day, the sun dispatches enormous energy for all the living being on earth, and in one-hour energy provided by the sun can supply enough power for a year. The photovoltaic effect is a conversion of energy harvested from the sun to electricity directly. This type of renewable energy is an attractive alternative to substitute fossil fuel since it is considered safer, no-pollution, reliable, maintenance free and long life-time about 20-30 years. PV system can be installed anywhere where the sunlight can reach it, and for a country with an abundance of sunlight, this is a perfect energy source. However, PV system also has a drawback, the sun only shines during daylight, therefore batteries are needed to support it, and batteries have short life-time compare to PV's life time, and they contribute to landfills and toxic waste. The fact is that not all of the sunbeam can be converted to electricity, only those with energy higher than bandgap energy can be converted. The material used to make PV module is also one of the factors affecting the efficiency of PV system [4].

Life cycle analysis (LCA) is an analysis to assess the environmental aspects and potential impacts related to a product by examining the inventory of inputs and outputs of the whole system, the environmental impacts related to those inputs and outputs and finally determine the sustainability of that product. LCA is taking the ISO 14040 as the standard for environmental analysis. This environmental analysis is taken for a life cycle time of a product or also known as cradle to grave, from material choice until disposal [5]-[28].

PV system as a new emerging renewable energy has to through an LCA analysis to get the whole idea of it, including its impacts to environmental, feasibility and sustainability. This paper discusses the LCA analysis of PV system from material choices, manufacturing process, implementation, and disposal or the afterlife. A cradle to gate and cradle to grave analysis will be conducted through a literature study. The result of this study will be an insight look of PV system application in Indonesia, from the best option of material choice, the method of application, energy payback time and the best method to dispose the module.

2. PV System Technology

PV system converts the incident from the sun into electricity to light up any loads. Solar cells are having the concepts of ideal diodes when exposed to light the current flow as forward bias. Figure 1.a shows the schematic diagram of a PV system. As the incident from the sun falls on a solar cell, photons coming from this incident bombard the n-type part of solar cells. The photons carry energies to p-type and give up those energies to electrons. Electrons having the new energies jump across the barrier into n-type. The movement of these electrons create charges of current and DC electricity is generated.

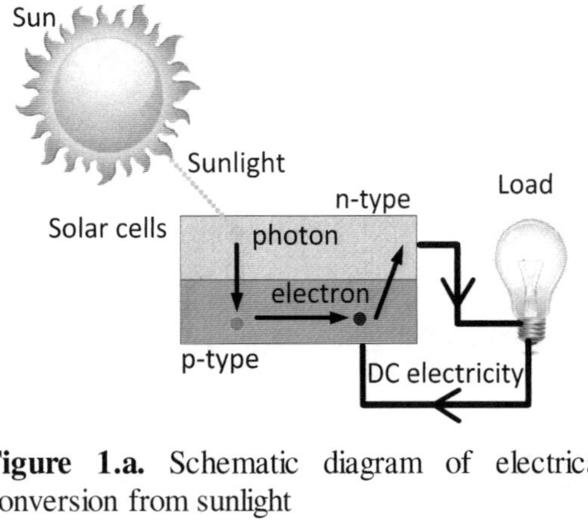

Figure 1.a. Schematic diagram of electrical conversion from sunlight

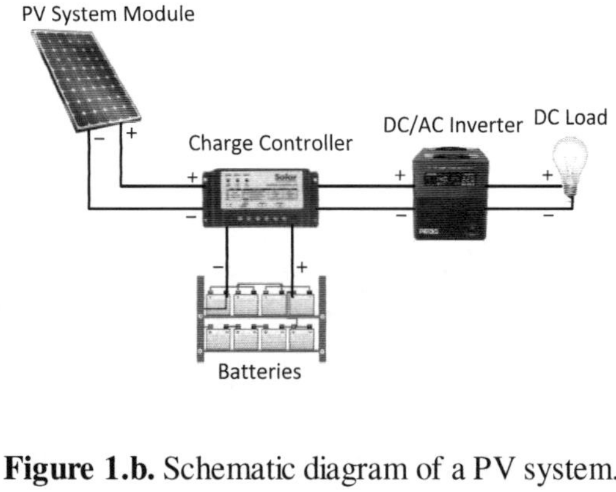

Figure 1.b. Schematic diagram of a PV system.

PV system produces DC electricity therefore in order to power up any devices requiring AC electricity, the system needs an inverter to convert from DC to AC electricity. One of the problems with electricity generation using PV system is the dependence on sunlight availability, and during the night or cloudy days, the supply of sunlight will not be enough to power up the load, therefore for standalone system, batteries are a necessary, and a charge controller is needed to ensure the system stops charging batteries once they are full. This basic requirement of standalone PV system is shown in Figure 1.b.

There are three types of PV system standalone system (off grid), grid-connected and hybrid. A standalone system is the type of PV system that does not depend on government electrical source, grid-connected is where the PV system is connected to government electrical source, and hybrid is where the system not only depends on PV system but also with other type electricity generator such as wind turbines.

3. LCA Methodology

LCA or life cycle analysis is a thorough assessment of a product evaluating its environmental impact based on ISO 14040 and sustainability prospects. LCA is a field of study, a technique, and also a specific study of a product. To assess a product, a preparation of data, model, and formula to use are a must, and also it is necessary to find out who to involve, to report and how to use the available data.

By conducting LCA, some benefits are gained such as to structure a large complex data, to compare the product with other alternative products, and at the same time to ease the process of benchmarking. LCA analysis is including cradle to grave, cradle to gate, and cradle to cradle in which considering the best of options for the product from the raw material up to after the life-time is over and disposed of.

Figure 2 shows the block diagram of LCA analysis of a PV system. The cradle to grave assessment is starting from the choice of raw material (cradle to gate), the process of producing the PV modules, transporting PV module to location, building, and installation, during the application of PV modules and finally dismantling and disposal. When PV modules' life-time is over, a choice will be made which of those components/materials that can be recycled or simply to be disposed of. LCA analysis is also including inventory analysis aim to evaluate how much materials used or produced that can affect environment. LCA analysis calculates energy used during all the process, as shown in Figure 2. where E_{mtl} is energy consumed during preparing the raw material, E_{dstr} is energy used in distribution, E_{istl} is energy used during installation, E_{use} is energy gained in a PV module life-time application, E_{mnc} is energy needed for maintenance of the system, E_{dsp} is energy used in dismantling and disposing, and E_{rcly} is energy used if recycling is possible.

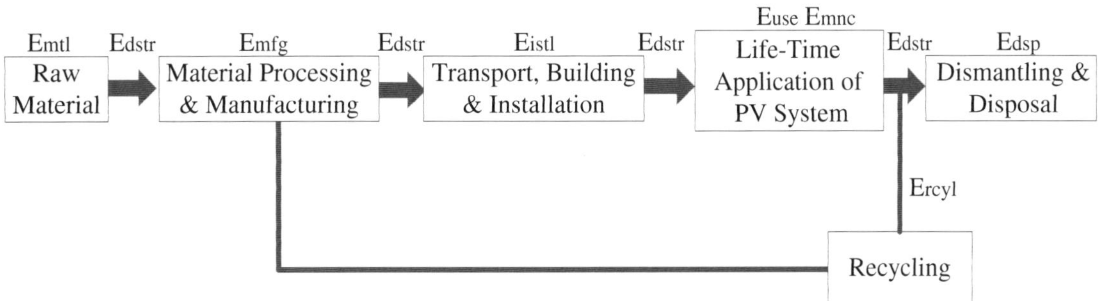

Figure 2. PV System LCA Scheme.

Impact assessment of all the process in Figure 2 is divided into classification, characterization, and weighting. All materials that are the influences to the environment are categorized in classification. In characterization, the amounts of output material are calculated by characterizing them to decide the indicator categories of the product impact. Weighting is to show the evaluation of damages caused and the estimation of environmental category [15]

The LCA assessment of PV system shown in Figure 2 will be explained in detail including the energy used and resulted and finally the recommendation of the best-applied PV module for Indonesia.

4. Comparison of PV Module Materials

PV technology for PV module materials now is up to the fourth generation. The first generation is the traditional technology involving a thick crystalline film made of mainly silicon, the monocrystalline or polycrystalline, and in another form is nanocrystalline silicon (nc-Si) [4][18].

The second generation is trying to improve the electricity generated by using thin film technology. This thin film technology not only is utilizing silicon but another type of semiconductor with better performance such as indium, gallium and phosphide. The common thin film technology material are cadmium telluride (CdTe), amorphous silicon (a-Si), and copper indium gallium diselenide ($Cu(In,Ga)Se_2$, CIGS). By stacking more than one type of material, it is expected more photo energy from the sunlight can be used to generate electricity. Stacking material type means creating different bandgap possibilities, as the only photon with energy larger than PV material energy bandgap can create electron movement that leads to charge movement and electricity generation.

The third generation is trying to overcome all the weakness in first and second generation called the Shockley-Quisser single bandgap limit and producing lower cost thin material for PV module. The third generation refers to PV technology using organic material, dye-synthesized, and quantum solar cell. The fourth generation is the technology to generate more energy by improving optoelectronic properties of the low-cost thin film PV. The third and fourth generation is still laboratory scale and not commercially produced in large scale.

Although silicon still is the most expensive type of solar cell material, however, it has the highest efficiency up to now for PV module in the market. Not only most expensive one, during the process of manufacturing, monocrystalline and polycrystalline require the highest energy compare to the second and third generation. The third and fourth generation are still in laboratory success yet.

From [4], the comparison output of different PV modules technology is given in Table 1

Table 1. Comparison of different PV modules technology

PV Module Technology		Material			
		c-Si	CIS	a-Si/Nc-Si	CdTe
Rated power	(W_P)	345	140	114	92.5
Rated current	(A)	6.02	2.98	2.18	1.94
Rated voltage	(V)	57.3	47	55	47.7
Short circuit current	(A)	6.39	3.31	2.6	2.11
Open circuit voltage	(V)	68.2	61.5	71	60.5
Dimensions	(m×m)	1.56×1.05	1.6×0.67	1.2×1.00	1.2×0.60
Expected life-time	(years)	25	25	25	25

5. Transport, Building, and Installation

Inventory analysis for transport means to calculate the energy needed to take PV modules from the factory to installation site. The distance from factory to installation site can be retrieved from Google Maps™.

Building and installation are including assembling balance of system (BOS), given it is a roof-top or a ground-mounted, and installing the modules to the system

As shown in Figure 1, PV system is not only PV modules but with other necessary devices such as inverter and battery. The transport, building, and installation of those devices should also be calculated.

The total energy required for the preparation of transport, building, and installation of a PV system is 84% for PV modules, 11% for Transportation, 3% for BOS, 2% for installation and 1% for inverter [25].

6. Maintenance

The good thing about PV system is that it requires minimum maintenance. PV system can last up to 25-30 years. Maintenance is including cleaning the surface of modules and checking the system performance periodically, other than that there is no special treatment.

Although PV modules can last up to 30 years, the battery is only up to 5 years, and other electronic devices have their own life-time. Therefore, maintenance is also including checking those devices and replacing them as required. The dependence to a battery can be cut by installing grid-connected PV system. In this type of PV system, temporary storage is not needed, if the produced electricity is not enough to power up the load, the system will connect automatically on grid and resume using PV system when the supply is back to normal.

7. Energy Payback Time

Energy payback time (in years) is the period needed for PV system to generate the same amount of energy with the energy used to produce the system and consumed throughout PV life-cycle [15]. From Figure 2, Energy payback time is given by

$$EPBT = \frac{Total\ Energy\ Consumed\ (kWH)}{Annual\ Power\ Generation\ \left(\frac{kWH}{Year}\right)} = \left(\frac{E_{mtl} + E_{mfg} + E_{istl} + E_{dsp} + Total\ E_{dstr}}{E_{use} - E_{mnc}}\right) \tag{1}$$

where E_{mtl} is energy consumed during the preparation of raw material, E_{mfg} is energy consumed during manufacturing process, E_{istl} is energy used during installation, E_{dsp} is energy used in dismantling and disposing, $TotalE_{dstr}$ is total energy used in distribution, E_{use} is energy gained in a PV module life-time application, and E_{mnc} is energy needed for maintenance of the system.

The CO_2 emission rate (g CO_2/kWh) is defined as an index to determine the effectiveness of a PV system related to environmental impact or global warming. CO_2 payback can be calculated by

$$CO_2 Emission\ Rate = \frac{Total\ Emission\ in\ a\ Life\ Cycle\ (g\ CO_2)}{Annual\ Power\ Generation\ \left(\frac{kWH}{Year}\right) \times Lifetime(year)} \tag{2}$$

The air pollutants and greenhouse gas emission emitted by PV system are nitrogen oxides, sulfur okside(SO_2), carbon monoxide (CO), PM_{10}, lead (Pb), and hydrocarbon (HC) for the air pollutant and green house emission are carbon dioxide (CO_2) and methane (CH_4).

According to [12], the EPBT of monocrystalline panels given by Table 2.

Table 2. EPBT of PV modules technologies [12]

References	Technology	Installation	EPBT (years)	Efficiency (%)	Performance Ratio
Alsema (2000)	Multicrystalline	Rooftop	2.50	13	0.75
	Monocrystalline	Rooftop	3.10	14	0.75

Fthenakis and Alsema (2006)	Multicrystalline	Rooftop	2.70	13.2	0.75
	Monocrystalline	Rooftop	3.10	14	0.75
Meijer et al (2003)	Multicrystalline	Not reported	3.50	14.5	0.75
Jungbluth (2005)	Multicrystalline	Rooftop	3.0-6.0	14.8	0.75
Fthenakis et al. (2011)	Multicrystalline	Rooftop	1.70	13.2	0.75
Fthenakis et al. (2009)	Monocrystalline	Rooftop	1.70	14	0.75
Fthenakis and Alsema (2006)	CdTe	Ground-mounted	0.80	9	0.80
Jungbluth et al (2008)	CdTe	Rooftop	2.70	7.1	0.75

8. PV Modules Disposal

The first large installation of PV system was in the beginning of 1990s, therefore a significant number of PV system ends its life time is starting on 2015. The end of life of PV modules leads to two choices to recycle or to let them all be land-filled wastes. The choice should consider how much energy achieved and consumed during the process. Recycling PV panels are not easy since they are assembled from multiple materials including glass (on the front cover of PV panels), aluminium (frame), synthetic materials to encapsulate and seal the silicon cells such as ethylene-vinyl acetate (EVA), polyvinyl butyral (PVB) and/or polyvinyl fluoride), silicon or other material, and metals such as lead, gallium, copper, and cadmium.

German and Belgian are among the first that started to recycle their end-life PV modules. The main recycled are modules made of monocrystalline, polycrystalline and amorphous silicon.

The recycle process started from dismantling the BOS by thermal step, the plastic, glass and framed parts are removed, dissembled, and sorted accordingly. The second step is to separate wafer from PV modules structure using etching process. The broken PV modules are cleaned to get the pure Si back. The Si obtained from used PV modules is re-melted into a polycrystalline ingot and produce new wafers.

One of the recycling processes is called Chevetogne from the Belgian system; this process is considered as the manual separation. In the US, PV Company uses an automatic recycle system where all the modules are broken down by the shredder, and by using a hammer mill those shredded materials are formed in 4–5 mm size. By using acid, all the semiconductor layers are removed, and other larger parts such as glass are separated. This process will end in a resulted filtered cake, and cut into new wafers [29][30].

Table 3. Life-time PV component as shown in Figure 2

No	Device	Life-time
1	PV modules	25 – 30 years
2.	Charge Controller	5 years
3.	Battery	5 years (average all types of batteries)
4.	Inverter	15 year
5.	Structure / Balance of System	25-30 years

From Table 3, the devices for PV system have a long life-time accept for battery and its charge controller. The battery contains a number of toxic chemical and heavy metals and without proper treatment, it can be a soil contamination if it is disposed of as a land-filled. Battery can be recycled, however, the application of battery can be avoided by applying grid-connected PV system.

IOP Conf. Series: Earth and Environmental Science **124** (2018) 012005 doi:10.1088/1755-1315/124/1/012005

9. The Feasibility of PV System Application in Indonesia

Indonesia is an archipelago country situated in equator (0.7893^{O} S, 113.9213^{O} N). As a tropical country, it is blessed with an abundance of sunlight all years long with the average of radiation intensity per day is 4.5kWh/m2, as shown in Figure 3 [31]. This blessing gives 112.000 GWp of solar energy potential, however up to now, it is only around 10 MWp utilized for powering electricity in Indonesia. Most of the Indonesian populations are distributed in remote and separate islands, about 60%, and grid supply electricity still cannot reach most of them.

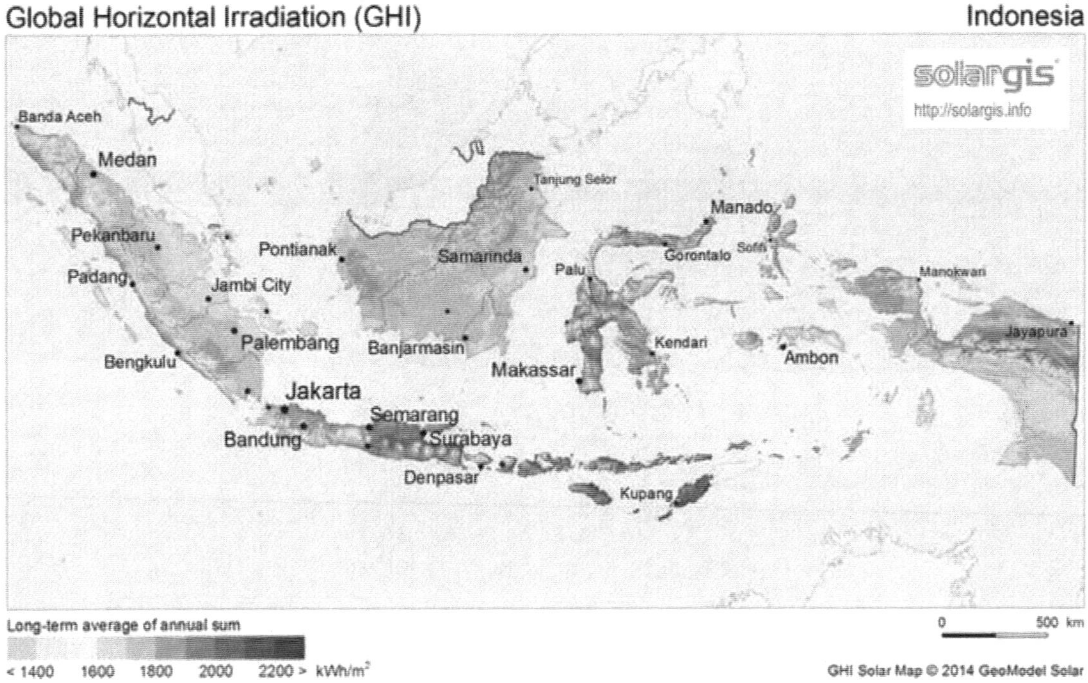

Figure 3. Sunlight incidents mapping in Indonesia [31].

PV system is one of the best options for electricity in Indonesia, not only in the cities but also up to remote areas (islands). Indonesian governments had come with several projects in proving PV system for the people living far from grid-connection. Altari Energi Surya, Mambruk Energy Internationa, and Guna Electro are private companies from Indonesia helping the government in installing 2239, 240 and 80 SHS (solar home systems) for the projects. SHS system consists of a PV panel with the capacity of 50 Wp. However, these projects, SHS was not working optimally, mainly due to most of the people were not ready for the new technology, and in the end went back to the old conventional way for energy (lighting) such as kerosene lamps [32].

However, with all those limitations in implementing PV system, Indonesian government and people have to start to find the alternatives to substitute fossil fuel that will soon diminish.

Based on LCA of PV system above, the most eligible material for PV system is monocrystalline due to having the highest efficiency, the fastest energy payback time and its availability in the market. For installation, the rooftop is the best method for cities in Indonesia to minimize shading, and also not requires additional space, however, this method only works for the new building, for more than 5 years building, ground mounted is better.

For remote area, the off-grid system is a must, however people have to be informed of battery management, and the solution has to be presented to ensure the battery replacement in every 5 years. For the cities application, grid-connected will be better, since the application of battery can be removed and PV system price can be reduced. Figure 4 shows grid-connected PV system in Indonesia.

IOP Conf. Series: Earth and Environmental Science 124 (2018) 012005 doi:10.1088/1755-1315/124/1/012005

The current applications of PV system in the rural area in Indonesia are including PV refrigerator system for vaccine storage in a rural clinic, and lighthouse navigation aid [33]. While in the cities, most application of PV system is off-grid and hybrid, tandem with a diesel generator.

The disposal of PV modules in Indonesia up to know is to be a land-filled waste since most of the PV modules are still imported, therefore the technology to re-process the used PV modules is not available yet. However, the long-life of PV modules is up to 30 years, therefore by that time the PV modules can be recycled into the new water and reused as mentioned in section 8.

Figure 4. Ground-mounted PV panels in Indonesia.

Despite the limitations above, PV system application in Indonesia keeps growing up, and more houses/companies install it. Indonesian government along with State-Owned enterprises promote the application of PV system on their buildings/company location, one of them is shown in Figure 4.

10. Conclusion

PV system is one of the choices for renewable energy, the substitute for diminishing fossil fuel. One of the main issues of renewable energy is its sustainability. LCA is a method to analyze the system thoroughly, starting from materials choice up to disposal, or called cradle to grave analysis. This method let us know how beneficial the system, comparing from the energy produced and energy consumed for the whole life-time. After looking closely the LCA of PV system (literature review), we can compare those conditions to the condition in Indonesia, the feasibility of PV system application in Indonesia is confirmed. PV system is one of the solutions for island country where the populations are spread among islands. Although, the current application is not so beneficial since the rural people are not well informed to understand the system, and see the long-term benefit of PV system. Fortunately, PV panels prices are getting lower, people are more aware of the need of clean renewable energy. In long-term, PV system will be installed more in Indonesia, as it has been a successful energy alternative in developing countries. The current most suitable material choice for PV system is monocrystalline for its highest efficiency among others, and shortest energy payback time. In 20 years, it is expected that Indonesia is able to recycle the used PV modules into a new wafer and can be re-used.

References

[1] Chen F F, An Indispensable Truth: How Fusion Power Can Save the Planet, 43 DOI 10.1007/978-1-4419-7820-2_2, © Springer Science+Business Media, LLC 2011
[2] U.S. Department of Energy The history of solar Office of Energy Efficiency \& Renewable Energy, accessed August 9, 2015.
[3] Dewi T, Risma P, Oktarina Y, Roseno M T, Yudha H M, Handayani AS, and Wijanarko Y, A Survey on Solar Cell; The Role of Solar Cell in Robotics and Robotics Application in Solar Cell Industry in Proc Forum in Research, Science, and Technology FIRST 2016 pp C19-C22.
[4] Jäger K, Isabella O, Smets A H M, van Swaaij R A.C.M.M, and Zeman M, 2014 Solar Energy: Fundamentals, Technology, and Systems, *Delft University of Technology.*
[5] Akarslan F 2012 Photovoltaic Systems and Applications *Modeling and Optimization of Renewable Energy System ed Sencan Arsu* chapter 2 pp 21-52.
[6] Turconi R, Boldrin A, and Thomas A, 2013 Life Cycle Assessment (LCA) of Electricity Generation Technologies: Overview, Comparability, and Limitation *Renewable and Sustainable Energy Review,* vol 28 pp. 555-565.
[7] Bekkelund K 2013 *Life Cycle Assessment of Thin Film Solar Panels* Master Thesis, Norwegian University of Science and Technology Norway.
[8] Beccali M, Cellura M, Longo S, and Mugnier D 2015 A Simplified LCA Tool for Solar Heating and Cooling System, *Proc Int. Conf. on Solar Heating and Cooling for Buildings and Industry (Istanbul)* vol 91 (Energy Procedia/Elsevier) p 317-324.
[9] Kommalapati R, Kadiyala A, Shahriar Md T, and Huque Z 2017 Review of the Life Cycle Greenhouse Gas Emissions from Different Photovoltaic and Concentrating Solar Power Electricity Generation System *Energies* vol 10 pp. 350-368.
[10] Keoleian G A and McD Lewis G 1997 Application of Life-cycle Energy Analysis to Photovoltaic Module Design *Progress in Photovoltaics: Research and Applications (Electronic Materials* vol 5*)* Broader Perspectives pp 287-300.
[11] Environment Canada 2012 *Assessment of the Environmental Performance of Solar Photovoltaic Technologies* Minister of the Environment Canada.
[12] Fthenakis V M, Kim H C, and Alsema Erik 2008 Emission from Photovoltaic Life Cycles *Environ. Sci. Technol.* 10.1021/es071763q.
[13] Anctil A and Fthenakis V 2012 Life Cycle Assessment of Organic Photovoltaic *Third Generation Photovoltaics* ed Fthenakis V (InTech) chapter 4 pp 91-110.
[14] Gerbiner S, Belboom S, and Léonard A 2014 Life Cycle Analysis (LCA) of Photovoltaic Panels: A Review *Renewable and Sustainable Energy Review* vol 38 pp 747-753.
[15] Ito M Life Cycle 2011 Assessment of PV Systems *Crystalline Silicon – Properties and Uses* ed Basu S (InTech) chapter 13 pp 297-312.
[16] Jungbluth N, Tuchschmid M, and de Wild-Scholten M 2008 Life Cycle Assessment of Photovoltaics: Update of Ecoinvent Data v2.0 *ESU-services Ltd.*
[17] Mousseau S 2011 *Life Cycle Analysis of the AFRESH Home Photovoltaic System* Bachelor of Engineering Thesis British Columbia Institute of Technology Canada.
[18] Meijer A, Huijbregts M A, Schermer JJ, and Reijnders L 2003 Life-cycle Assessment of Photovoltaic Modules: Comparison of mc-Si, InGap and InGap/mc-Si Solar Modules *Progress in Photovoltaics: Research and Applications (Electronic Materials* vol 5*)* Broader Perspectives vol 11 pp 275-287.
[19] NREL 2012 Life Cycle Greenhouse Gas Emission from Solar Photovoltaic *downloaded from www.nrel.gov/harmonization.*
[20] Baharwani V, Meena N, Dubey A, Brighu U, and Mathur J, 2014 Life Cycle Inventory and Assessment of different Solar Photovoltaic System, *Proc Power and Energy System: Toward Sustainable Energy (PESTSE 2014).*
[21] Fthenakis V, Kim H.C., Frischknecht R, Raugei M, Sinha P, and Stucki M, 2011 Life Cycle Inventories and Life Cycle Assessment of Photovoltaic Systems, *International Energy Agency (IEA) PVPS Task 12*, Report T12-02:2011

[22] R. Frischknecht, R. Itten, P. Sinha, M. de Wild-Scholten, J. Zhang, V. Fthenakis, H. C.Kim, M. Raugei, M. Stucki, 2015, Life Cycle Inventories and Life Cycle Assessment of Photovoltaic Systems, *International Energy Agency (IEA) PVPS Task 12*, Report T12-04:2015.

[23] Galagan Y and Andriessen R 2012 Organic Photovoltaics: Technologies and Manufacturing *Third Generation Photovoltaics* ed Fthenakis V (InTech) chapter 3 pp 61-90.

[24] Frankl P and Masini A 1997 Simplified Life-cycle Analysis of PV Systems in Buildings: Present Situation and Future Trends *Center for the Management of Environmental Resources.*

[25] Sumper A, Robledo-Garcia M, Villafáfila-Robles R, Bergas-Jané J ,and André-Peiró J, 2011 Life-cycle Assessment of a Photovoltaic System in Catalonia (Spain) *Renewable and Sustainable Energy Review,* vol 15 pp 3888-3896.

[26] Alsema, E. A. Energy Pay-back Time and CO2 Emissions of PV Systems. *Prog. Photovolt: Res. Appl.* **2000**, *8*, 17–25.

[27] Jungbluth, N. 2005 Life cycle assessment of crystalline photovoltaics in the Swiss ecoinvent database.*Prog.Photovolt:Res.Appl*, vol.13, No. 4, pp. 429–446.

[28] Murphy F, and McDonnel K 2017 A Feasibility Assessment of Photovoltaic Power Systems in Ireland; a Case Study for the Dublin Region, *Sustainability*, vol 9 pp. 302-316.

[29] Bine Informationsdienst, Recycling Photovoltaic Modules, ed Hahne A and Hirn G, FIZ Karlsruhe, Germany.

[30] Fthenakis V M, 2000 End-of-life Management and Recycling of PV Modules, *Energy Policy* vol 28, pp 1051-1058.

[31] https://janaloka.com/potensi-energi-matahari-di-indonesia/ accessed on Sunday, 06 August 2017.

[32] Retnanestri M, Outhred H, Healy S, 2005 Off-Grid Photovoltaic Applications in Indonesia: An Assessment of Current Experience, *The University of New South Wales*.

[33] Adra N, Reuze B, and Sudrajat A, 2009 State of the Art of Solar Energy Applications in Indonesia, RENDEV.

Near Zero Energy House (NZEH) Design Optimization to Improve Life Cycle Cost Performance Using Genetic Algorithm

Y Latief[1] , M A Berawi[1] , A B Koesalamwardi[1] , L S R Supriadi[1]

[1]Civil Engineering Department, Faculty of Engineering, Universitas Indonesia, Depok, Jawa Barat, 16424

Abstract. Near Zero Energy House (NZEH) is a housing building that provides energy efficiency by using renewable energy technologies and passive house design. Currently, the costs for NZEH are quite expensive due to the high costs of the equipment and materials for solar panel, insulation, fenestration and other renewable energy technology. Therefore, a study to obtain the optimum design of a NZEH is necessary. The aim of the optimum design is achieving an economical life cycle cost performance of the NZEH. One of the optimization methods that could be utilized is Genetic Algorithm. It provides the method to obtain the optimum design based on the combinations of NZEH variable designs. This paper discusses the study to identify the optimum design of a NZEH that provides an optimum life cycle cost performance using Genetic Algorithm. In this study, an experiment through extensive design simulations of a one-level house model was conducted. As a result, the study provide the optimum design from combinations of NZEH variable designs, which are building orientation, window to wall ratio, and glazing types that would maximize the energy generated by photovoltaic panel. Hence, the design would support an optimum life cycle cost performance of the house.

Keywords: Near Zero Energy House, Optimum Design, Genetic Algorithm, Life Cycle Cost

1. Introduction

Energy efficiency of a building becomes one of the key element to reduce carbondioxide emition and to save energy. Improving energy efficiency is a more affordable way to fight climate change as well as increasing energy availability. Energy conservation in residential sector gain interest in order to earth preservation. Net Zero Energy Building (NZEB) is a Zero Energy House or Building which energy is much reduced through efficiency where the consumed energy was retrieved from renewable sources. As for NZEH (Near Zero Energy House), it is a residential building which consume energy less than energy generated by itself. By harvesting energy from the sun, NZEH building will consume energy its produced and minimize energy from outside [1].

Indonesia is an equatorial country which get solar exposure along the year. This is very potential for renewable energy as well as NZEH. However, this potential is not yet maximize due to the high price of solar generated electricity technology [2]. This high price of solar-generated electricity for residential also because it is not integrated to central electricity network and not yet optimized.

High efficiency building like NZEH should be financially feasible. Good strategy and solution is needed to reduce energy consumed for the building an continuous provide renewable energy [3]. This problem also creates a contradiction since low energy building uses more expensive materials which directly increase the construction cost [4]. This needs compromise and optimization of green technology and construction cost so that low energy building can be used by all parties and viewed to be feasible based on its life cycle cost. In terms of optimization, genetic algorithm (GA) is part of an evolutionary algorithm which has been proven as a tool to solve this type of problem [5]. Therefore, the objective of this study is to determine the NZEH design with the optimum cost using genetic algorithm method.

2. Literature Review

The basic concept of Zero Energy Building (ZEB) is a building that can produce its energy from clean, renewable and less expensive sources. Another definition showed that ZEB can produce on-site energy enough or more than its annual energy consumption. A zero energy building can be defined in several ways depend on its objective. For example, building owners usually consider more on cost. There are four different definition which: net zero site energy, net zero source energy, net zero energy cost and net zero energy emmisions[1].

Near Zero Energy House technology is usually more expensive than conventional building. Those technologies used are natural ligthning, solar PV, wind turbine, light emitting diode (LED) lamp, automatic HVAC system, etc. However, NZEH technology can preserve more energy as well as less operational and maintenance cost. Therefore, it is important to use proper analysis to measure cost reduction during the building life cycle [6].

Some key features of building life cycle cost (LCC) are [7]:

a. using economical concept;

b. considering total cost of ownership and anticipated cash flow of building system;

c. calculate time value of money and cashflow;

d. Including tangible and intangible assets such as health, productivity to be quantified in monetary terms.

In calculating the LCC of a building, the cost components that need to be analyzed consist of [7]:

1. Initial cost – the cost spent before utilizing the building and its facilities. Example: PV panel installation cost, energy saving light installation, and others.

2. Operational cost – the cost spent during the building operational years. This cost is usually spent annually, not including maintenance and repair cost. Example: electricity cost, taxes, insurance, etc.

3. Maintenance Cost – the cost that is already scheduled regularly for maintenance system. This cost is needed to prolong the building usage.

4. Replacement Cost – the cost needed for replacing main components of the building which is necessary to sustain the operational system of the building.

5. Salvage Value – the value of the a system in the end period of the LCC analysis.

6. Disposal cost – the cost needed to dispose a system component which is already obsolete. As an example, the mercury contained in CFL and HID light could contaminate the ground water, hence it should be dispose to the waste treatment plant.

According to Wang et al (2006) [8], the life cycle cost of a NZEH is influenced by the initial construction cost and its annual operational cost. The initial construction cost is influenced by the renewable energy technology chosen for the NZEH, the choice of building envelope, and the building utilities. As for the building operational cost is affected by the performance and efficiency of the renewable energy technology used and the utilities energy consumption [9].

Genetic Algorithm:

Genetic algorithm (GA) is a numerical technique to solve both constrained and unconstrained optimization problems which is based on natural selection approach, the process that refers to biological evolution. It iteratively modifies a population of individual solutions. It started with searching by selecting random sample and operated a stochastic approach to find the optimum solution of the problem [10] [11].

The basic principle of GA: A solution of a problem is an individual and a group of individuals which is called a population. A generation is a new population. In GA binary, each individual is represented by a binary string that is called a chromosome, where it codes all of the important parameters of that specific individual. Each individual has a specific objective function. GA provides the process of controling the evolution from various generations. There are three basic GA numerical operators, which are: reproduction, cross-over and mutation, which will be applied to each selected individual / chromosome. These operators are based on the nature of biological evolution which are converted numerically. All of these numerical operators would provide the possible optimized solution based on the selected individual.

A GA process starts with creating several possible solution which will be named as the first population (parent). This population is evaluated and operated using the basic GA operator. The result will provide a new population which consist of better evolutionary combination of individuals than the previous population. The new population should be evaluated by the user whether it provides the optimum solution. This process can be iterated based on the needs and the complexity of the problem. This process is illustrated in Figure 1 [12] [5].

IOP Conf. Series: Earth and Environmental Science **124** (2018) 012006 doi:10.1088/1755-1315/124/1/012006

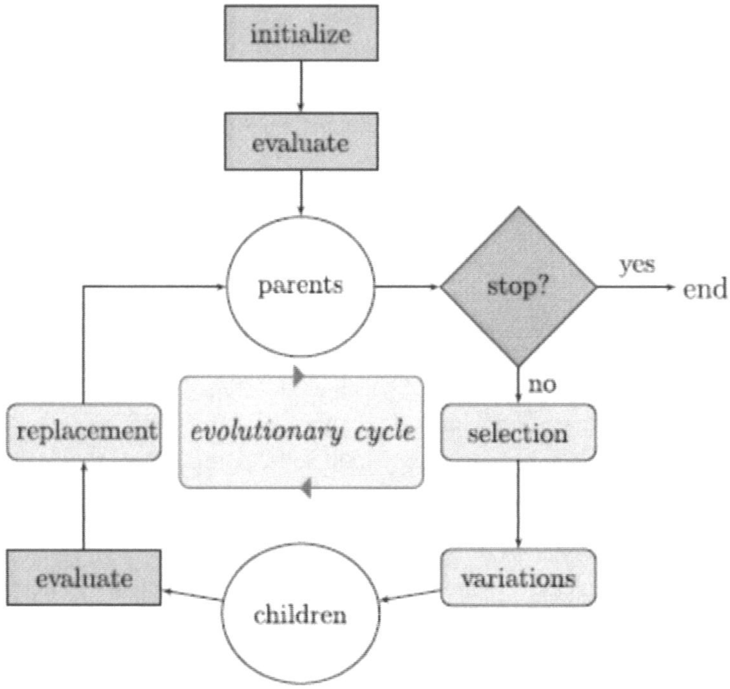

Figure 1. Optimization cycle using Genetic Algorithm [12]

The application of GA is relatively easy to understand, where it adopts biological evolution process as the basic operator. GA also uses codes, mostly binary codes, which are also easy to implement [5]. Figure 2 shows an example of the binary code used in GA operator.

Figure 2. Design variable combination using Genetic Algorithm [12]

3. Methodology

In order to obtain the objective, which is to is to determine the NZEH design with the most optimum cost using genetic algorithm method, the study conducted an experiment as the research strategy. The study firstly determine the design variables for NZEH that will be optimized using literature studies. Secondly, the study conducted the experiment using a software for building performance simulation (EnergyPlus software) with the genetic algorithm as the main rules of the optimization process.

For the experiment, the study designed one schematic model of a house (using software Sketchup and Open Studio) that is used as the sample for optimization. The model was converted into the data file for EnergyPlus software. The data, supported with various weather database, was simulated in EnergyPlus using the GA rules. Here are the details of the sample and input data for the simulation process:

IOP Conf. Series: Earth and Environmental Science **124** (2018) 012006 doi:10.1088/1755-1315/124/1/012006

The sample is a one-storey house (Figure 3.) with the total area of 90m^2. The location of the sample was in Bekasi, West Java, Indonesia, which means the weather and climate data for the sample are based on this location. The sample was occupied by 4 persons. For the experiment, there were 2 models that will be used, which were one sample with PV panel and other NZEH design variables and the other sample without PV panel and referred as a conventional house.

Figure 3. Front view of the sample

Other data for experiments were: weather and day – night temperature of the selected location; various building material specifications; occupancy schedule for the sample; financial parameters of LCC analysis (example: initial cost, operational and maintenance cost, and other related cost) – these costs were benchmarking costs from the typical building of the sample.

The design variables for the experiment were building orientation (azimuth), PV panel, window-wall ratio (WWR) and glazing [12]. The detailed descriptions of the design variables that were simulated in EnergyPlus is shown in Table 1.

Table 1. Design variables for experiment

No.	Sub-Variables	Code	Number of variation
1	Azimuth	Azi	8
2	PV Panel	PV	2
3	WWR	WWR	5
4	Glazing	GT	5

The rules of the simulation combined the three basic GA operators: reproduction, cross over dan mutation of the individual (each individual has 4 sub-variables from Table 1.)

The simulation process would find the optimized design variables that would minimize the net energy consumption from the sample. The energy simulation was based on the PV panel energy data and the household energy consumption data (daytime and night). The output from the experiment would be the data of net energy consumption from the sample of varied design variables and the Net Present Value (NPV) of the sample (PV sample and non PV sample), based on the LCC analysis. During the experiment, checklists and EnergyPlus software were used as the research instruments.

4. Results and Discussion

The NZEH design variables that were used as the parameters for the study was determined by literature studies. According to Boeck et al (2013) [13] and Bucking et al (2013) [12], there are six main design variables to be used for NZEH, which are: building orientation (azimuth), building envelope (glazing), fenestration (WWR), HVAC (Heating, Ventilation, and Air Conditioning), PV panel, and lighting. In this study, the experiment only used four of the variables (azimuth, WWR, glazing dan PV panel) due to the one model sample. This study only focused on how the renewable energy technology, in this case the PV panel, would perform when various parameters were simulated.

The experiment was firstly started with a pilot simulation. The pilot simulation showed how the sample would fit in the simulation process. The result showed that schematic model of the sample comply with the EnergyPlus simulation process, and therefore could proceed to the main simulation process. The main experiment simulated 400 variations of the 2 samples (1 sample with PV and 1 sample non-PV (conventional)) and results are discussed below:

Building azimuth – The building orientation has an effect towards the building's net energy consumption. This is due to the sun direction, where a house / building that faces directly towards the sun, then the house would save more energy during the day (due to the sunlight). However, it could increase the energy consumption at night due to the high room temperature. The result of the experiment showed that from the 2 samples, there were different trends of energy consumption towards the azimuth direction (shown in Figure 4 and Figure 5.).

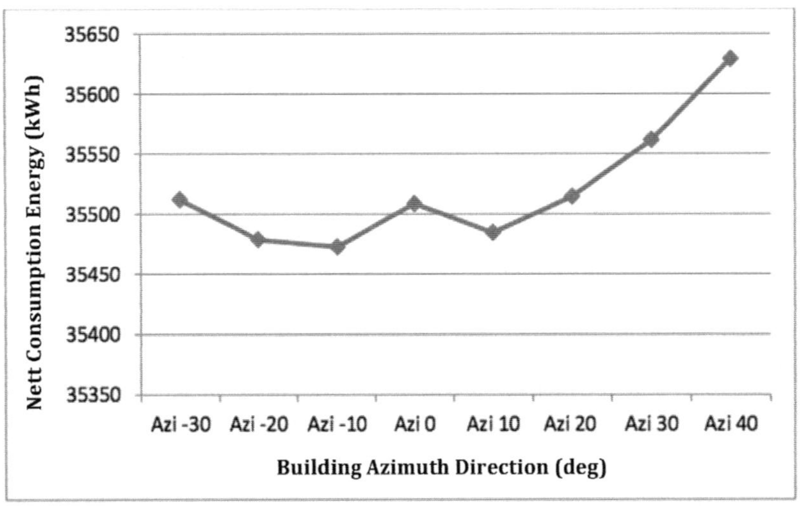

Figure 4. Building azimuth vs Net Energy Consumption for NZEH sample

IOP Conf. Series: Earth and Environmental Science **124** (2018) 012006 doi:10.1088/1755-1315/124/1/012006

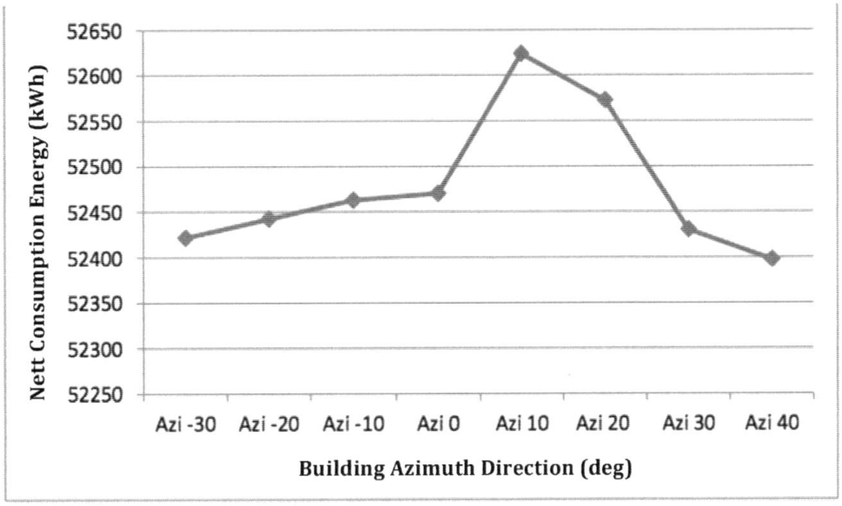

Figure 5. Building azimuth vs Net Energy Consumption for conventional sample

From the figures, it was found that the highest sun radiation point was at -10 degree to 10 degree of the azimuth. For the conventional sample (non-PV), the high sun radiation were causing higher energy consumption for the use of air conditioning. As for the PV sample (NZEH sample), the high sun radiation were causing higher energy produced by the PV panel and not consumed by the sample.

Window-to-wall Ratio (WWR) – Figure 6 and Figure 7 showed that there were an increase of energy consumption of the 2 samples in line with the increase of the ratio between the window area and the wall facade. This is due to a higher sun radiation that went through the house.

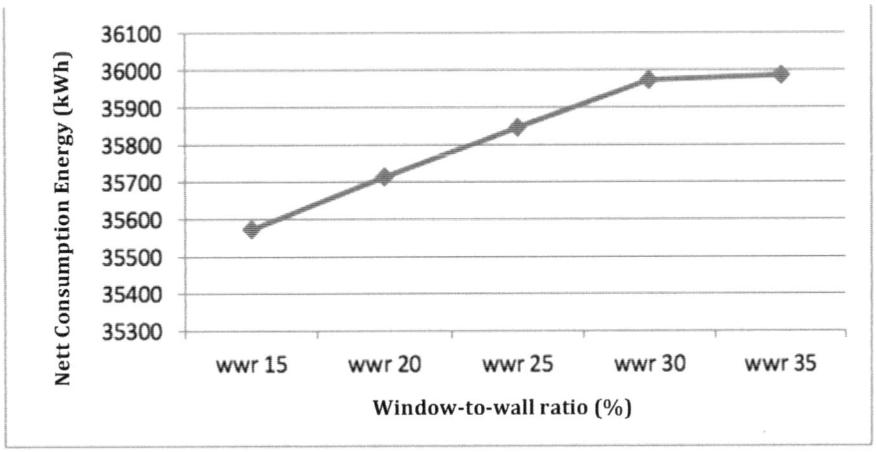

Figure 6. WWR vs Net Energy Consumption for NZEH sample

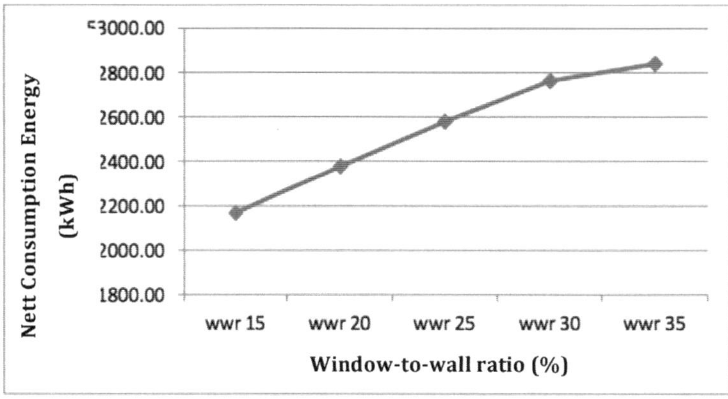

Figure 7. WWR vs Net Energy Consumption for conventional sample

Glazing type – For housing in the tropical climate, glazing on the window is needed to prevent a much higher sun radiation that would enter the house. This could also be useful to reduce the use of mechanical air conditioning which would consume more energy. From the experiment, blue-color glazing provide the least energy consumption for both samples, as shown in Figure 8 and Figure 9.

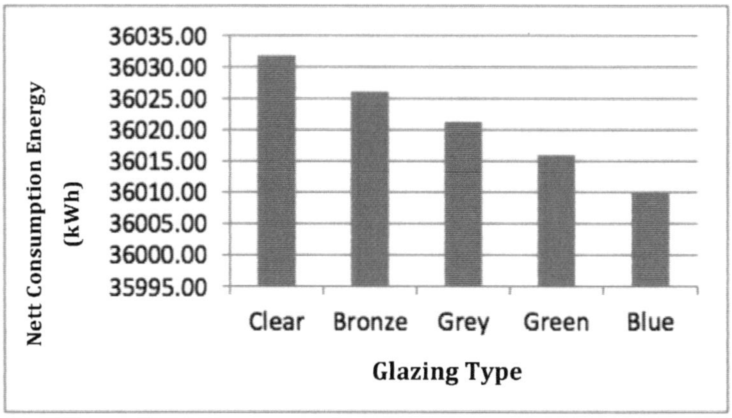

Figure 8. Glazing type vs Net Energy Consumption for NZEH sample

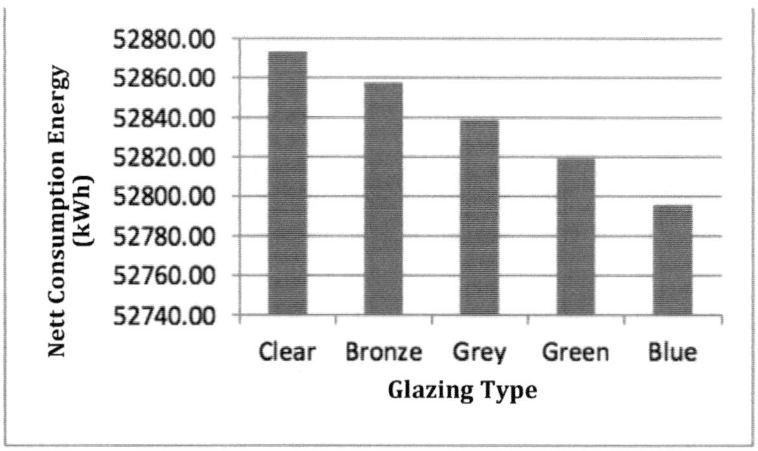

Figure 9. Glazing type vs Net Energy Consumption for conventional sample

Net energy consumption and NPV results

Based on the 400 variations of the 2 samples, Figure 10 showed the data distribution from the sample with PV (NZEH sample) – named Zone B, and the non-PV sample (conventional sample) – named Zone A.

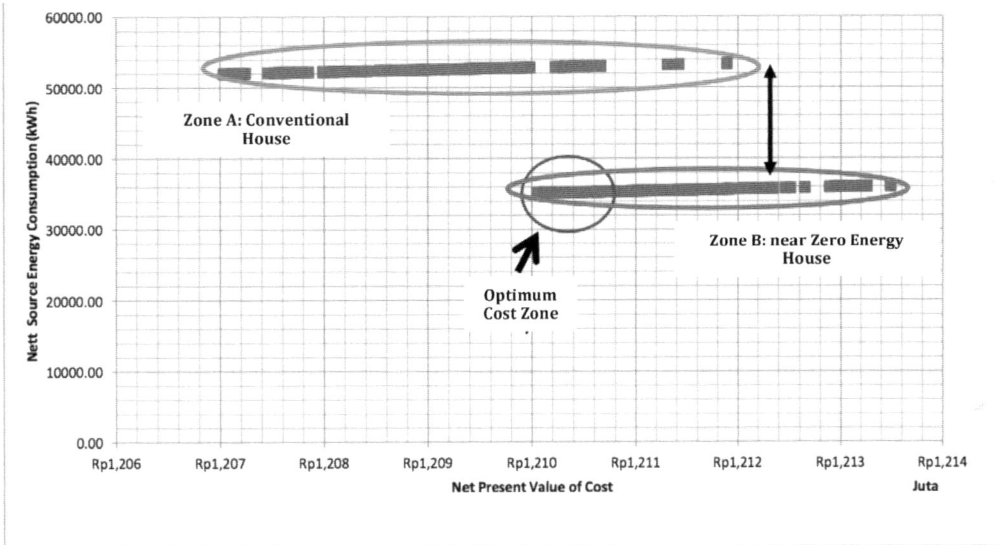

Figure 10. Data distribution of NZEH and Conventional samples on NPV and Net Energy Consumption

From the graph, it showed that by using PV panel, a house with 4 persons occupancy with normal electricity usage could save net energy consumption up to 32.2%. Some of the Zone B designs had lower NPV due to the different variation of design variables. From the graph, it also showed that the optimum design was the NZEH designs with the lower NPV than the conventional house. Also, the optimum cost was found at Zone B, with the lowest energy consumption. Some of the conventional samples had higher NPV than the NZEH samples. This is due to the high operational cost that came from the high electricity utilities costs for air conditioning. Lastly, based on this graph, the optimum design for NZEH (with the optimum LCC) was: a sample with azimuth -10 degree with PV panel, WWR of 15% and 20%, and the use of blue-color glazing for window.

5. Conclusion

To conclude, this study determine the NZEH design with the optimum cost, which uses the NZEH design variables such as PV panel, azimuth (building orientation), fenestration (window-to-wall ratio), and glazing for windows. The design variables were combined and simulated using Genetic Algorithm numerical rules with the help of building performance simulation software. The result showed that a NZEH with azimuth of -10 degree with PV panel, WWR of 15% and 20% and the use of blue-color glazing provides the optimum life cycle cost (an optimum NPV of the house) and optimum net energy consumption. This finding can be a reference to design a typical NZEH, particularly in tropical climate.

This study has its limitation, where it only focuses on a one model house as a sample, and only using 4 design variables. For further studies, the object can be a larger house or building, with different building materials that could be combined. Also, optimization methods other than Genetic Algorithm (such as direct search, sequential search, etc) can also be adopted to find newer approaches in optimizing the NZEH designs.

6. Acknowledgement

The authors would like to acknowledge Research Grant from Ministry of Research, Technology and Higher Education Republic of Indonesia (Kemenristekdikti) for the funding support of the research project (*Hibah Penelitian Unggulan Perguruan Tinggi 2017*).

7. References

[1] Torcellini, P., Pless, S. & Deru, M., 2006. *Zero Energy Buildings: A Critical Look at The Definition.* Pacific Grove, California, National Renewable Energy Laboratory.

[2] Dhany, R. R., 2013. *Detik Finance.* [Online] Available at: http://finance.detik.com/read/2013/02/22/192004/2177424/1034/ini-alasan-pembangkit-listrik-ma tahari-sulit-berkembang-di-ri [Diakses 17 September 2013].

[3] Milan, C., Bojesen, C. & Nielsen, M. P., 2012. A Cost Optimization Model for 100% renewable Residential Energy Supply Systems. *Energy,* Issue 48, pp. 118-127.

[4] Milajic, A., Milosevic, P. & Belkovic, D., 2013. Optimal Green Building Design Using Multi-Objective Genetic Algorithm. *Reporting For Sustainability,* pp. 119-123.

[5] Caldas, L. & Norfolk, L., 2003. Genetic Algorithms For Optimization Of Building Envelopes And The Design And Control of HVAC Systems. *Journal Solution Energy Engineering,* pp. 343-351.

[6] Hartungi, R. & Pye, C., 2009. *Life Cycle Cost (LCC) for Sustainable Building,* Lancashire, UK: Univesity of Central Lancashire.

[7] Blank, L. & Tarquin, A., 2005. *Engineering Economy.* 6th edition. New York: McGraw- Hill.

[8] Wang, W., Rivard, H. & Zmeureanu, R., 2006. Floor Shape Optimization for Green Building Design. *Advanced Engineering Informatics,* Issue 20, pp. 363-378.

[9] Marszal, A. J. et al., 2010. Zero Energy Buildings - A Review of Definitions and Calculation Methodologies. *Energy and Buildings,* Issue 43, pp. 971-979.

[10] Goldberg, D., 1989. Genetic Algorithms in Search Optimization and Machine Learning. Addison: Wesley.

[11] Holland, J., 1975. Adaptation in Natural and Artificial Systems. Michigan: Michigan Press.

[12] Bucking, S., Zmeureanu, R. & Athienitis, A., 2013. An Information Driven Hybrid Evolutionary Algorithm For Optimal Design Of A Net Zero Energy House. *Solar Energy,* Issue 96, pp. 128-139.

[13] Boeck, L. D., Audenaert, A. & Mesmaeker, L. D., 2013. Improving The Energy Performance Of Residential Buildings: A Literature Review. *HUBRUSSEL RESEARCH PAPER.*

Optimizing the Utility Power of a Geothermal Power Plant using Variable Frequency Drive (VFD) (Case Study: Sibayak Geothermal Power Plant)

R H M Sinaga[1] and **Y Manik**[2]

[1,2]Engineering Management Study Program, Del Institute of Technology, Laguboti, Toba Samosir 22381, Indonesia

E-mail: rizal.sinaga@del.ac.id

Abstract. Sibayak Geothermal Power Plant (SGPP) is one of the plants being developed by Pertamina Geothermal Energy (PGE) at the upstream phase. At the downstream phase, State-owned Electricity Company (PLN) through PT. Dizamatra Powerindo is the developer. The gross capacity of the power plant is 13.3 MW, consisting 1 unit of Monoblock (2 MW) developed by PGE and 2 units (2×5.65 MW) operated through Energy Sales Contract by PLN. During the development phase of a geothermal power plant, there is a chance to reduce the utility power in order to increase the overall plant efficiency. Reducing the utility power can be attempted by utilizing the wet bulb temperature fluctuation. In this study, a modeling process is developed by using Engineering Equation Solver (EES) software version 9.430. The possibility of energy saving is indicated by condenser pressure changes as a result of wet bulb temperature fluctuation. The result of this study indicates that the change of condenser pressure is about 50.8% on the constant liquid/gas (L/G) condition of the wet bulb temperature of 15°C to 25°C. Further result indicates that in this power plant, Cooling Tower Fan (CTF) is the facility that has the greatest utility load, followed by Hot Well Pump (HWP). The saving of the greatest utility load is applied trough Variable Frequency Drive (VFD) instrumentation. The result of this modeling has been validated by actual operations data (log sheet). The developed model has also been reviewed trough Specific Steam Consumption (SSC), resulting that constant L/G condition allows the optimum condition on of the wet bulb temperature of 15°C to 25°C.

Keywords: Geothermal, Sibayak, variable frequency drive, wet bulb

1. Introduction

Initial design of geothermal power plant is started on choosing suitable thermodynamics cycle, preparing power plant modeling, and then determining technical specification for every equipment component through simulation [9]. This procedure is conducted in order to optimize the net output of a power plant by setting up the thermodynamics parameters. During its operation, power plant systems should be analysed for both full and partial load. Partial load operation is conducted to know power plant system condition during some equipment maintenance without shutting down the power plant. A Simulator is needed to make some decisions for such condition.

[8] studied the performance of both full and partial load of Lumut Balai geothermal power plant with capacity 2× 55 MW using Engineering Equation Solver (EES) software. In the study, the simulation of partial load using both wet bulb temperature and Non-Condensable Gas (NCG) percentage were used as input variable parameter. The study shows that power plant performance is more significantly influenced by wet bulb temperature fluctuation rather than NCG percentage.

Wet bulb temperature fluctuation makes a chance for optimizing power plant system by reducing energy use from certain equipment. Variable Frequency Drive (VFD) is applied for both Main Cooling

Water Pump (MCWP) and Cooling Tower Fan (CTF) [3]. The installation of VFD to Hot Well Pump (HWP) in Darajat Geothermal Power Plant can be applied starts from wet bulb 15.5°C until 18.5°C. the highest gain of VFD HWP is obtained when the wet bulb temperature 18.5°C, about 0.80% compared to constant Liquid/Gas (L/G) [6].

This study aims to optimize the overall plant efficiency of Sibayak Geothermal Power Plant (SGPP) through modeling and simulating the wet bulb temperature fluctuation with particular attention to the refrigeration system.

2. Sibayak Geothermal Power Plant (SGPP)

SGPP is one of the plants being developed by Pertamina Geothermal Energy (PGE) at the upstream phase. At the downstream phase, State-owned Electricity Company (PLN) through PT. Dizamatra Powerindo is the developer. SGPP is located at 3°10'12.7" N and 98°29'17.9"E about 65 km southwest from Medan, capital city of North Sumatera. The gross capacity of the power plant is 13.3 MW, consisting 1 unit of Monoblock (2 MW) developed by PGE and 2 units of single flash system (2×5.65 MW) operated through Energy Sales Contract by PLN.

SGPP is situated on Singkut Caldera about 1400 up to 2200 masl. It is surrounded by three active volcanoes: Pintau Mt, Sibayak Mt, and Pratektekan Mt. SGPP exploration field was begun by Pertamina from 1989 to 1991. Initial exploration well was located between Sibayak Mt. and Pratektekan Mt. Well drillilng was held by Pertamina since 1991 until 1997 consisted of 7 production wells (SBY-1, SBY-3, SBY-4, SBY-5, SBY-6, SBY-7, SBY-8) and 3 reinjection wells (SBY-2, SBY-9, and SBY -10).

Reservoir's temperatures are ranging between 240 and 300°C, and the pressure is about 45 bar-g. Reservoir's enthalpy is up to 1100 kJ/kg and the mass flow rate from well about 51.4 kg/s and pressure 9 bar [7]. Through simulation using EES, [7] obtained that the increase of turbine output power is in tandem with the decrease of condenser pressure. The increase of turbine output power is also related to the increase of utility load. As the result, lower condenser pressure will reduce turbine output power for constant steam supply. Through modeling, optimum output turbine (about 20.858 MW) for 2 units power plant is obtained when separator, condenser, and inlet turbine pressure are on 8 bara, 0.08 bara and 7.7 bar, respectively. However, based on actual condition, turbine and condenser pressure is operated on 7.5-8.0 and 0.1 - 0.3 bara, respectively.

3. Theoretical Basics

3.1. Power Plant Efficiency

The higher NCG percentage, the lower the efficiency of power plant system because NCG will affect the condenser pressure. Condenser pressure is maintained by the increase of NCG extraction. Gas extraction system needs motive steam (for steam jet ejector) or utility power (for vacuum pump) which is increased as well as NCG percentage increase [1]. Through developed simulator, performance of SGPP is analyzed both by net power and steam supply. The performance is called Specific Steam Consumption (SSC) [8] which defines the amount of steam that is needed to generate 1 MWh electricity.

SGPP is developed by single flash power plant, whose flow process through *T-s* diagram is shown in figure 1[2]. Based on figure 1, the process starts from well head (point 1). Fluid undergoes pressure decrease and isenthalpic $(h_1 = h_2)$ inside the separator (point 2). Steam from separator flows to demister and enters to the turbine (point 3). While brine from separator (point 4) and demister (point 7) flows to cooling tower before injected through reinjection well. Turbine expansion process is shown from point 3 to 5. Point 3 to *5s* shows isentropic process inside turbine $(s_3 = s_{5s})$. Therefore, steam quality at point *5s* is

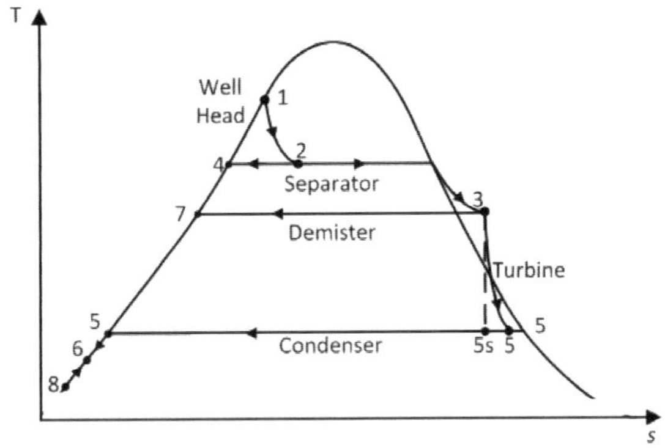

Figure 1. Single Flash Condensing Temperature (***T***) and entropy (***s***) Diagram

SGPP is developed by single flash power plant, whose flow process through *T-s* diagram is shown in

$$x_{5s} = \frac{s_{5s} - s_{5l}}{s_{5v} - s_{5l}} \quad (1)$$

Enthalpy at point h_{5s} :

$$h_{5s} = x_{5s} \cdot h_{5v} + \left(1 - x_{5s}\right) \cdot h_{5l} \quad (2)$$

Turbine isentropic efficiency is denoted as η_t. Turbine exit enthalpy (h_4) is calculated using:

$$\eta_t = \frac{h_3 - h_5}{h_3 - h_{5s}} \quad (3)$$

Thus, turbine output power is:

$$\dot{W}_t = \dot{m}_3 \cdot \left(h_3 - h_5\right) \quad (4)$$

Where $\dot{m}_3$ denotes steam flow rate from well head, $\dot{m}_3 = x_2 \cdot \dot{m}_1$

Exit steam turbine is condensed by cooling water from cooling tower (point 8). Condensate temperature is obtained by:

$$\dot{m}_8\left(h_6 - h_8\right) = \dot{m}_3\left(h_5 - h_6\right) \quad (5)$$

3.2. Variable Frequency Drive (VFD)

The lower wet bulb temperature for constant cooling tower operation, the lower the temperature of circulating cooling water. Thus, VFD should be applied to reduce utility power of Hot Well Pump (HWP). Meanwhile, VFD Cooling Tower Fan (CTF) is applied to maintain both cooling water temperature and mass rate while wet bulb fluctuating. Maintaining of both temperature and mass rate are controlled by the adjustment of cooling air mass rate [3]. This study analyzes plant optimizing by VFD CTF only.

Flowing capacity difference through a pump is controlled by an open valve or by a pump rotation arrangement. The capacity difference is shown in figure 2 where figure 2(a) shows the decreasing of capacity Q_1 to Q_2 by lowering of open valve. For this way, pump utility power remain constant. While figure 2(b) shows that decreasing capacity is controlled by the speed of pump rotation. This way can make the utility power difference. Optimization of HWP utility load is applied trough pump rotation diference.

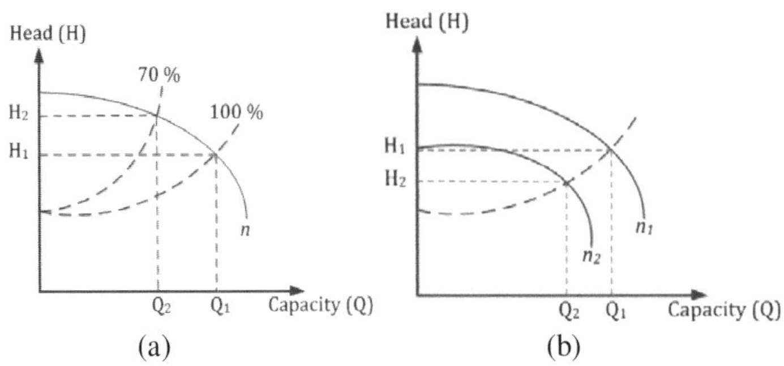

Figure 2. The Arrangement of Flowing Capacity in a Pump by (a) Open Valve, (b) Pump Rotation.
Note: Percentage denotes valve opening

4. Power Plant Modeling

As a part of the modeling stage, SGPP is described as a Process Flow Diagram (PFD) in figure 3. Red line shows steam flowing, blue line shows condensate and cooling water flowing. While purple line shows NCG flowing. This model is then validated with the log sheet data from unit 1 of the single flash of Sibayak Plant (5 MW) using heat and mass balance [5].

4.1. Steam Turbine

Steam flowing during turbine expansion is shown through stream number 4 to 6 in figure 3. Assumed that there is no heat transfer through turbine. Thus, work rate that turbine produced is:

$$\dot{W}_{turbine} = \dot{m}_4 \cdot (h_4 - h_6) \cdot \eta_{generator} \tag{6}$$

4.2. Condenser

Condensation process is maintained by cooling water spraying directly to steam exit turbine as shown in figure 3 through stream line 16. Using heat and mass balance in the condenser, equation (7), (8), and (9) are obtained.

- NCG mass balance : $\dot{m}_{NCG,6} = \dot{m}_{NCG,10}$ (7)
- Water and steam mass balance : $\dot{m}_{s,6} + \dot{m}_{w,15} + \dot{m}_{w,16} = \dot{m}_{w,7}$ (8)
- Energy balance :

$$(\dot{m}_{s,6} + \dot{m}_{NCG,6})h_{mix,6} + \dot{m}_{w,15} \cdot h_{w,15} + \dot{m}_{w,16} \cdot h_{w,16} = \dot{m}_{w,7} \cdot h_{w,7} + \dot{m}_{NCG,10} \cdot h_{NCG,10} \tag{9}$$

Equations (7) to (9) are used to obtain both condensate temperature and mass flow rate.

4.3. Cooling Tower

Cooling tower is designed to reduce condensate heat. It is processed by the decrease of condensate temperature through cooling air absorbing around cooling tower. Then cooling water will be injected and some of it circulated again for condensing process. Mechanical draft cooling tower is used in SGPP. Utility power of cooling tower fan is formulated in equation (10):

$$\dot{W}_{motorfan} = \frac{\dot{v}_{air} \cdot \Delta p}{\eta_{fan} \cdot \eta_{motorfan}} \tag{10}$$

Where $\dot{v}_{air}$ denotes air volumetric rate (m^3/s) and Δp as pressure drop inside cooling tower. Cooling process in cooling tower is influenced by wet bulb temperature and air relative humidity. Cooling water temperature that can be reached is higher than wet bulb temperature. The difference of those temperatures is called approach. While the difference between inlet and exit temperature of cooling water is called as range. The value of both range and approach depends on cooling tower design.

IOP Conf. Series: Earth and Environmental Science **124** (2018) 012007 doi:10.1088/1755-1315/124/1/012007

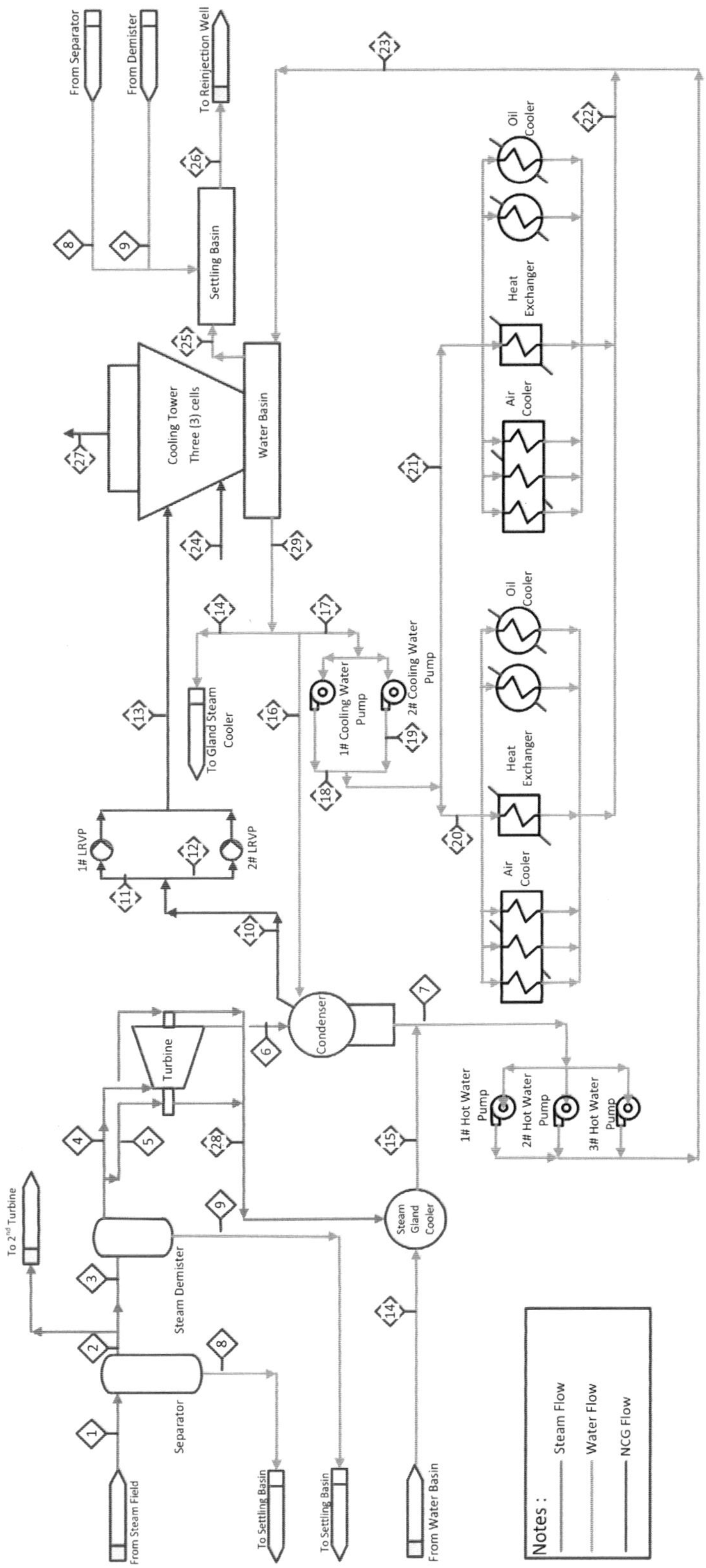

Figure 3. Process Flow Diagram (PFD) SGPP Unit I

The concept of cooling tower heat transfer is developed based on the difference on potential enthalpy between air and water [4]. It is assumed that each water particle is wrapped up by air surface. The difference enthalpy between air and water causes cooling process. While cooling process, heat is released by water is same as heat received by cooling air. Therefore,

$$\dot{m}_L \cdot c_p \cdot \left(T_{w,in} - T_{w,out}\right) = \dot{m}_a \cdot \left(h_{a,out} - h_{a,in}\right) \quad (11)$$

Where $\dot{m}_L$ and $\dot{m}_a$ define mass flow rate of water and air, respectively. Whereas, c_p, T, and h denote heat capacity, temperature, and enthalpy, respectively.

Cooling tower is modeled based on figure 3 which is flown by stream numbers 23, 24, 25, 27, and 29. It has 3 cells. The first and second fans have the same output power, each fan 75 kW while the third fan output is 37 kW.

Following equations are applied based on heat and mass balance in cooling tower:

- NCG mass balance : $\dot{m}_{NCG,13} = \dot{m}_{NCG,27}$ (12)
- Dry air mass balance: $\dot{m}_{a,24} = \dot{m}_{a,27} = \dot{m}_a$ (13)
- Wet air mass balance: $\omega_{24}\dot{m}_{a,24} + \dot{m}_{w,23} = \dot{m}_{w,25} + \dot{m}_{w,29} + \omega_{27}\dot{m}_{a,27}$ (14)
- Energy balance :

$$\omega_{24}h_{a,24} + \dot{m}_{w,23}\frac{h_{w,23}}{\dot{m}_a} = h_{a,27} - h_{a,24} + \omega_{27}h_{a,27} + \dot{m}_{w,29}\frac{h_{w,29}}{\dot{m}_a} + \dot{m}_{w,25}\frac{h_{w,25}}{\dot{m}_a} \quad (15)$$

Equation (13) to equation (15) provide cooling air mass rate $\left(\dot{m}_{a,24}\right)$ and cooling water mass rate $\left(\dot{m}_{w,25}\right)$, also L/G. Where L/G denotes to the ratio of condensate water mass rate $\left(\dot{m}_{w,23}\right)$ to cooling airmass rate $\left(\dot{m}_{a,24}\right)$, about 1.099. Variable ω and h defines specific humidity and enthalpy, respectively.

4.4. Gas Removal System

Liquid Ring Vacuum Pump (LRVP) is one of gas removal systems. It is used in Sibayak. The higher NCG concentration, the higher work rate of LRVP to remove the gas.

4.5. Pump

There are several main pumps used in Sibayak, those are HWP, Cooling Water Pump (CWP), and reinjection pump. Pump work rate is calculated by:

$$\dot{W}_{motorpump} = \frac{\dot{m}_{water} \cdot \dot{v}_{water} \cdot \Delta p}{\eta_{pump} \cdot \eta_{motorpump}} \quad (16)$$

Where $\dot{m}_{water}$ denotes water mass rate (kg/s), v as specific volume (m^3/kg), Δp as pressure difference (kPa), and η denotes the efficiency.

5. Result

5.1. Constant Liquid/Gas (L/G)

Estimation of cooling water temperature is obtained by cooling tower heat transfer theory. Both cooling water (T_{16}) and condensate temperature (T_7), for wet bulb temperature range from 15°C to 25°C, are displayed in figure 4(a). The higher wet bulb temperature, the higher both cooling water temperature. Cooling water temperature range from 18.64°C up to 26.94°C.

Meanwhile, condensate temperature is 30.66°C up to 39°C. Figure 4 (a) shows that the temperature difference of the two unit processes are relatively constant over the increasing wet bulb temperature. This phenomenon occurs because of the mass rate of the cooling water is relatively constant. The higher the temperature of cooling water, the higher the pressure of condenser. The increasing of

IOP Conf. Series: Earth and Environmental Science **124** (2018) 012007 doi:10.1088/1755-1315/124/1/012007

condenser pressure is shown in figure 4(b). For wet bulb range 15°C to 25°C, the increase of condenser pressure is 0.1357 – 0.2 bar, which is about 47.38%. The increase of condenser pressure will affect the gross power where the higher condenser pressure, the lower gross power will be, for constant inlet turbine pressure.

As wet bulb temperature increases, both gross and net power slightly decrease for constant steam mass flow rate. Figure 5 shows the change of gross and net power. Gross power decreases from 5.64 MW to 5.201 MW, about 7.78%. In addition to gross power decrease, utility power also decreases from 5.289 MW down to 4.856. Most of utility power is consumed both by cooling water pump and cooling tower fan. Figure 5 also shows the increase of SSC net as the wet bulb temperature increases. SSC net rises from 11.74 kg/MJ to 12.79 kg/MJ, about 9%.

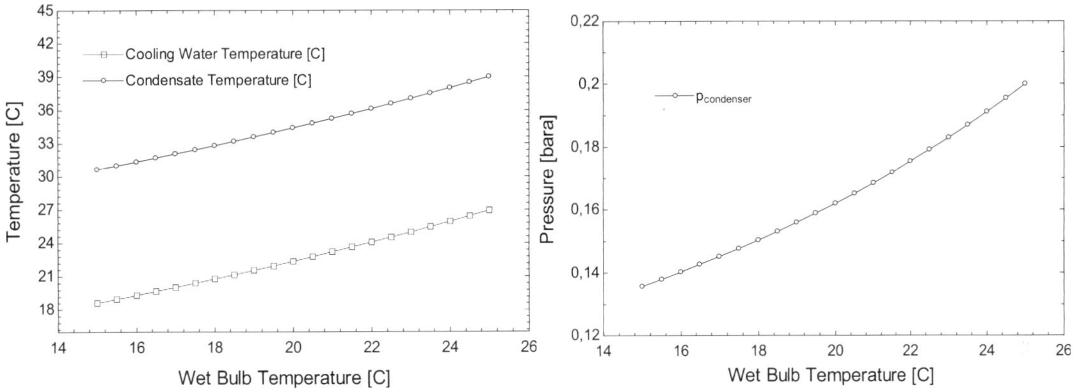

(a) (b)

Figure 4. (a) Cooling Water and Condensate Temperature on Constant L/G, (b) Condenser Pressure on Constant L/G

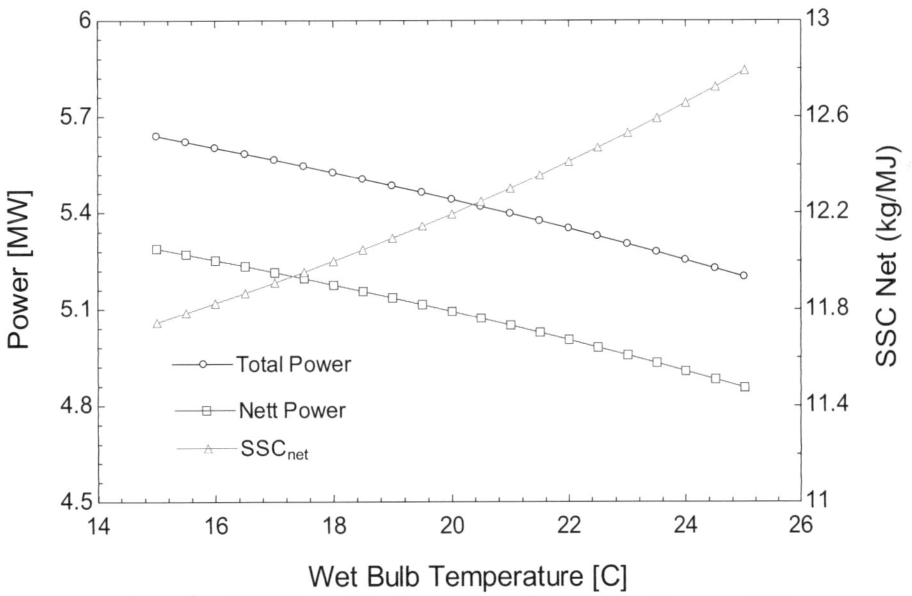

Figure 5. SSC Gross/Net Power and SSC Net on Constant L/G

5.2. VFD Application in Cooling Tower Fan (CTF)

The installation of VFD to cooling tower fan aims to keep constant condenser pressure, 0.2 bara. Constant condenser pressure is obtained by maintaining both cooling water temperature and mass rate. Figure 6(a) shows the effect of VFD installation to maintain cooling water temperature. Both cooling water and condensate temperature are maintained at 26.94°C and 39°C, respectively, as wet bulb temperature increases. VFD also aims to maintain cooling water mass rate, which is shown in figure 6 (b), as much as 568.2 kg/s. Meanwhile, cooling air is increased to maintain constant cooling water temperature and mass rate, in order to maintain constant condenser pressure. The increase is 238.9 kg/s up to 544.8 kg/s, about 56.15%

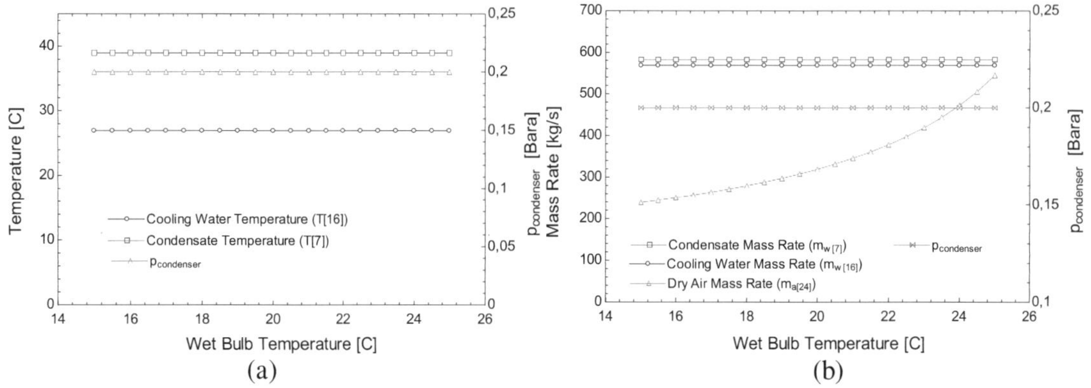

Figure 6. (a) Cooling Water & Condensate Temperature and Condenser Pressure on VFD CTF, (b) Cooling Water, Condensate, and Cooling Air Mass Rate on VFD CTF

Cooling air mass rate is adjusted by the number of cooling tower fan. Figure 7(a) illustrates the number of utilized fans. SGPP has three units of fan, consisting of two 75 kW fans (fan *b* and *c*) and a unit of 37.3 kW fan (fan *a*). In wet bulb temperature range 15 °C up to 20°C, both fan *a* and *b* should be utilized. Meanwhile, for wet bulb temperature 20°C - 25°C, all fans should be utilized simultaneously .

The impact of VFD installation is compared to constant L/G to find out optimization possibility. Figure 7(b) shows the difference of fan work rate of those two scenarios. The highest energy saving is obtained by VFD when wet bulb 15°C, that is 56.17%. Meanwhile when wet bulb temperature is 25°C, VFD does not provide any impact. Based on figure 7(b), VFD should be installed for wet bulb less than 25°C.

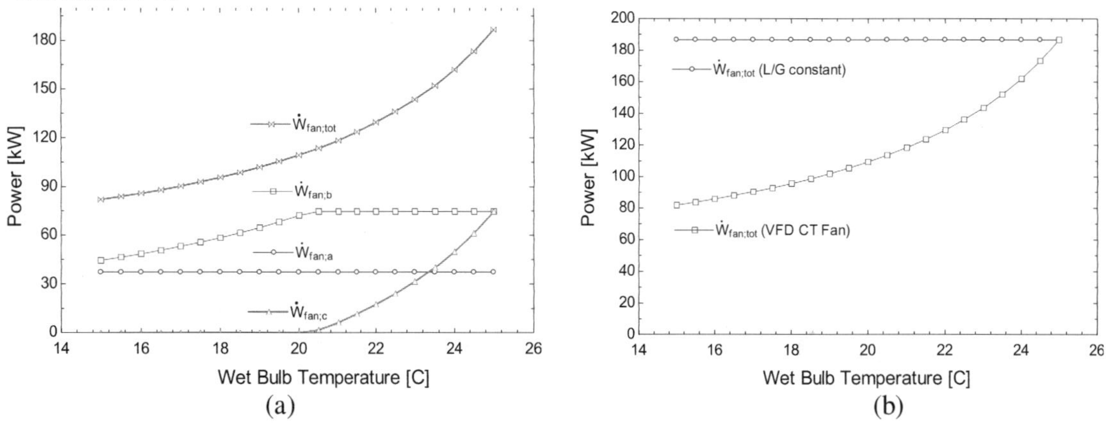

Figure 7. (a) Utility Fan Power on VFD CTF, (b) Utility Fan Power on Constant L/G and VFD CTF

The decision of the best scenario considers not only utility fan power but also gross and net power. Figure 8(a) displays that both net and gross power in constant L/G are higher than VFD CTF. Therefore, VFD does not affect the whole wet bulb temperature. Another consideration for choosing the best scenario is by evaluating the SSC net. SSC is defined as the amount of steam to generate 1 MWh net electricity. The less steam supply required to generate 1 MWh electricity, the more efficient the power plant is. SSC net consideration is displayed in figure 8(b). The highest SSC net difference is obtained when wet bulb equals 15°C, about 6.76%. As a final consideration, constant L/G operation is more efficient than VFD CTF operation for wet bulb temperature less than 25°C.

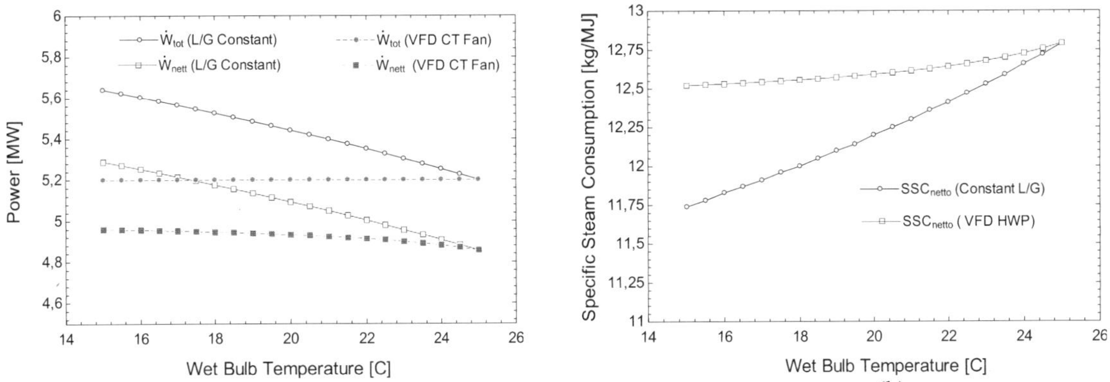

(a) (b)

Figure 8. (a) Gross and Net Power on Constant L/G and VFD CTF, (b) SSC between Constant L/G and VFD CTF

Due to the lack of data, the power plant performance can not be evaluated for wet bulb greater than 25°C. However, by figure 8(b), it is possible to assume that constant L/G SSC net is higher than VFD CTF operation for wet bulb greater than 25°C. For complete verification of this guess, need more complete data are required.

6. Conclusion

Having developed the model and simulated some variables, conclusions are obtained that the developed model of SGPP approaches the real condition based on validation of daily operational data (log sheet). The impact of VFD installation is compared to constant L/G to find out optimization possibility. Through the model developed, the highest fan utilization energy saving is obtained by VFD when wet bulb 15°C, that is 56.17%. However, the decision of the best scenario considers not only utility fan power, but also steam supply and net power which is formulized in term of SSC. The model shows that the SSC net of VFD is higher than constant L/G which concludes that constant L/G operation is more efficient. The model is developed in wet bulb temperature range 15°C up to 20°C where constant L/G is more efficient rather than VFD CTF. Constant L/G gives the highest saving 6.64 % when wet bulb 15°C.

References

[1] DiPippo R 2007 *Geothermal Power Plants : Principal, Applications, Case Studies, and Environmental Impact* (Darmouth, Massachusets: Elsevier Ltd.)
[2] DiPippo R 2015 *Geothermal Power Plants : Evolution and Performance Assessments* Geothermics
[3] Effendi H 2009 *Main Cooling Water System Optimization Kamojang Unit 2 Geothermal Power Plant, PT. Indonesia Power* (Bandung, ITB) pp 10-45
[4] Freeston, D 1996 *Geothermal Technology:Teaching The Teachers - Course Stage III.* pp 15-20
[5] Moran M J and Howard N Shaphiro 2011 *Fundamental of Engineering Thermodynamics 7th ed.* (John Willey and Sons) pp.154-180

[6] Sinaga R H M and Darmanto P S 2016 *Energy Optimization of Geothermal Power Plant (Case Study:Darajat Geothermal Field Unit III)*, IOP Conf. Series Earth and Environmental Science 42 (5th ITB International Geothermal Workshop (IIGW), Bandung) pp. 7-13
[7] Siregar P H 2004 *Optimization of Electrical Power Production Process for the Sibayak Gethermal Field, Indonesia* (Reykjavik, Iceland: The United Nations University) pp. 349-365.
[8] Sulistyardi H B 2010 *Basic Design of Lumut Balai 2x55 MW Geothermal Power Plant, Indonesia* (Geothermal Training Program, The United Nations University) pp. 614-627
[9] Swandaru R B 2009 *Modelling and Optimization of Possible Bottoming Units for General Single Flash Geothermal Power Plants*, M.Sc. (Thesis. Reykjavik, Iceland: The United Nations University) pp. 12-15

Acknowledgment
This study was made possible by funding support from Del Institute of Technology under the framework of Internal Research Grant. The authors are also supported by PT. Dizamtra Powerindo, especially Mr. Sofjan Hadi, by providing access to operational supporting data for this research.

IOP Conf. Series: Earth and Environmental Science **124** (2018) 012008 doi:10.1088/1755-1315/124/1/012008

Policy Model of Sustainable Infrastructure Development (Case Study : Bandarlampung City, Indonesia)

C Persada[1], S R P Sitorus[2], Marimin[2], and R D Djakapermana[3]
[1]University of Lampung, Bandarlampung, Lampung, Indonesia
[2]Graduate School, Bogor Agricultural University, IPB Campus Dramaga, Bogor, Indonesia
[3]Public Works Department, Republic of Indonesia
* E-mail of the corresponding author: citrapersada8@yahoo.com

Abstract. Infrastructure development does not only affect the economic aspect, but also social and environmental, those are the main dimensions of sustainable development. Many aspects and actors involved in urban infrastructure development requires a comprehensive and integrated policy towards sustainability. Therefore, it is necessary to formulate an infrastructure development policy that considers various dimensions of sustainable development. The main objective of this research is to formulate policy of sustainable infrastructure development. In this research, urban infrastructure covers transportation, water systems (drinking water, storm water, wastewater), green open spaces and solid waste. This research was conducted in Bandarlampung City. This study use a comprehensive modeling, namely the Multi Dimensional Scaling (MDS) with Rapid Appraisal of Infrastructure (Rapinfra), it uses of Analytic Network Process (ANP) and it uses system dynamics model. The findings of the MDS analysis showed that the status of Bandarlampung City infrastructure sustainability is less sustainable. The ANP analysis produces 8 main indicators of the most influential in the development of sustainable infrastructure. The system dynamics model offered 4 scenarios of sustainable urban infrastructure policy model. The best scenario was implemented into 3 policies consist of: the integrated infrastructure management, the population control, and the local economy development.
Keywords: ANP, city, dynamic model, policy, Rapinfra, sustainability infrastructure

1. Introduction

High population growth in city areas has implications for the improvement of the community infrastructure needs. The relationship between cities and infrastructure is now emerging as a key city policy issue [14]. Many relevant aspects and actors involved in city infrastructure development and planning and it required a comprehensive and integrated policy to be sustainable [7, 14, 17, 27]. A variety of strategies, policies, plans and programs of action for the development of an integrated and sustainable infrastructure in urban have been prepared, but the development of urban infrastructure still faces unresolved issues [16, 28]. Infrastructure development does not only affect the economic aspects, but also social and environmental aspects, those are the main dimensions of sustainable development. Therefore, it is important to determine the measuring instrument to identify the ability to build sustainable infrastructure.

Previous studies on sustainable infrastructure reflected the need to design and manage engineering systems by the environment, social and economics consideration. The study include: municipal water system sustainability criteria [6, 23], sustainable transportation [5, 9, 10, 12, 23, 30], drinking water system [6, 23, 25], waste water systems [6 23 26], storm water systems [2, 4, 19, 29], green infrastructure [1, 15, 22] and solid waste [3, 24]. Based on these studies it is known that there has been no research on criteria and indicators for integrated and sustainable infrastructure. The main objective of this research is to formulate policy of sustainable infrastructure development, with specific objectives are: firstly, to measure the level of sustainability of the city's infrastructure; secondly, to formulate the indicators that influence the sustainability of city infrastructure development and thirdly, to design a policy model of sustainable urban infrastructure with dynamic system. The influential indicators was obtained from the results of stakeholders assessment, public opinion and the assessment of infratsructure planning report against all indicators of sustainable infrastructure development. This research was conducted in Bandarlampung which is one of the fast-growing large cities in Indonesia and in year 2015 it is expected becoming a metropolitan city [21]

2. The Research Method

The scope of the city infrastructure research restricted to a basic network infrastructure that influence city development, namely: transportation, water systems (drinking water, storm water, waste water), green open spaces and solid waste. The research was carried out by Multi Dimensional Scaling (MDS) method, which consisted of the application of Rapinfra (Rapid Appraisal of Infrastructure), Analytic Network Process (ANP) and dynamic system model. The primary data were the data obtained directly from respondents through Focus Group Discussion (FGD) and the data collected from questionaire survey of the community in Bandarlampung City. FGDs conducted in Bandarlampung 3 times (July 2013, August 2013, and January 2014). The sampling technique in this research was the method expert survey [13] by conducting in-depth interviews to the 15 respondents who had been determined. For the survey to the public, the sampling technique the purposive sampling was used to 126 respondents. Analyzing the sustainability status using Multi Dimensional Scaling (MDS) with software Rap-fish [8] were modified to Rapinfra. Sustainability status in this study were analyzed with the five dimensions of sustainability namely environmental, social, economic, technology and good governance. Sustainability analysis conducted through three stages: 1) Attributes determination for sustainable infrastructure development, which includes dimensions of environmental, economic, social, technology and good governance. 2) The valuation of each attribute in an ordinal scale based on sustainability criteria for each dimension. The scoring is based on the result of questionaires in accordance with the stipulated requirement. The scores ranged from 0 – 3, which is interpreted from strongly disagree (poor) to strongly agree (good). 3) Results of the scoring was analyzed using Rapinfra program to determine the position of the sustainability status in each of these dimensions (Table 1).

Leverage analysis was used to determine the sensitive attributes which was very influential in improving the status of sustainable infrastructure development. The determination of sensitive attributes was based on the priority of analysis leverage result that taking into account of the changes the root mean square (RMS) ordination on the X axis. The greater the change in RMS value, the greater the role of these attributes in increasing the sustainability status of city infrastructre.

Table 1. Sustainability Index and Status

Index	Category
0,00 – 25,00	Poor (not sustainable)
25,01 – 50,00	Less (less sustainable)
50,01 – 75,00	Fair (fairly sustainable)
75,01 – 100,00	Good (Sustainable)

Analytic Network Process (ANP) was used to determine the influential indicators of sustainable infrastructure development. The steps of selecting influential indicators as follows: 1). Determination of criteria and indicators based on expert consultation from the results of the previous analysis was based on a literature study, stakeholders and public opinion 2). Determination of the relationship between indicators was obtained through questionnaires 3). Construction of an alternative network model was based on the results of step 1 and 2. 4). Scaling interest for alternative indicators of sustainable infrastructure development. 5). Testing consistency of pairwise comparison matrices that already meet the inconsistency ratio ≤ 10%. The next step is to calculate the weights of criteria and synthesis of indicators alternative of sustainable infrastructure development with a super decisions software use. Dynamic system is a comprehensive and integrated way of thinking, it is able to simplify the complex issues without losing the important things of concern [18]. A system dynamics model can not only arrange and describe the complicated connections among each element in different levels, but also deal with dynamic processes with feedback in a system. It also can predict the complex system change under different scenarios, which is very useful in examining and recommending policy decisions in city management.

3. Results and Discussion

3.1. The level of sustainability infrastructure

The findings on sustainable criteria and indicator for infrastructure development from various studies were summarized in Table Appendix 1. Analyzing for indicators of sustainable infrastructure development on previous research, there were obtained 5 criteria with 50 indicators for sustainable infrastructure development (Table 2). The criteria and indicators which resulted from literature review in Table 1 were used for further consulted with experts through focus group discussions (FGD). From the FGD, it fixed the number of criteria to be 5 criteria, while the number of indicators was reduced to 47 indicators.

Table 2 Criteria and indicators of sustainable infrastructure

Environmental Criteria	Social Criteria	Economic Criteria	Technology Criteria	Good Governance Criteria
1. Land carrying capacity	1. Population growth	1. Economic growth	1. Drainage systems	1. Regulation
2. Conservation area damage growth	2. Number of poor	2. City Revenue growth	2. Sewage system	2. Planning (sectoral)
3. Built up area growth	3. Human Development Index (HDI)	3. Investment growth	3. Drinking water system	3. Inter- sector institution
4. Slum area growth	4. The community sewage system	4. The city budget growth	4. Solid waste management	4. The visionary leadership
5. Air quality	5. Catchment areas by public	5. Level of per capita income	5. Green Open Space systems	5. Spatial planning
6. Water quality	6. Processing trash by community	6. Minimum city wage	6. Road systems	6. Law enforcement
7. Land quality	7. Artesian/shallow wells by community	7. Levels of local economic growth	7. Bicycle lanes / non- motorcycle vehicle	7. Socio-political conditions
8. Availability of water resources	8. Levels of security &safety	8. Infrastructure services fee	8. Facilities for pedestrians	8. Call center
9. Traffic congestion level	9. Unemployment rate	9. Land value	9. Public transportation	9. Budgeting
10. City landscape	10. Levels of traffic accident			10. Human resource capacity in goverment
	11. Community behaviour (culture)			11. Community participation

The results of MDS using Rapinfra showed that the sustainability index value of environmental criteria was 42.88% as shown in Figure 1. It was classified as less sustainable, due to 2 attributes laid in bad score which were the rate of conservation, damage and level of traffic congestion. Seven (7) attributes laid in moderate score which were land carrying capacity, the growth rate of built up area, slum area growth, air quality, water quality, land quality and water resources. The less sustainable status was influenced by 4 key indicators that leverage analysis results and it can be seen in figures root mean square (RMS). Key indicators were indicators of the middle to the highest RMS value. The RMS of key indicators were air quality level; the rate of conservation area damage; the level of water quality; the soil quality level (Figure 2).

Other chart which is the output of MDS with Rapinfra sustainability analysis for the economic criteria, social criteria, technological criteria and governance criteria similar such as output over the environmental dimension. The sustainability index value for social criteria was 15.80 % and classified as not sustainable.

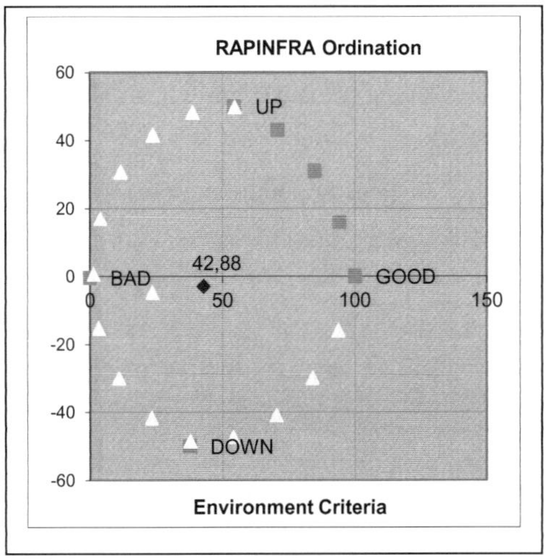

Figure 1. Value Criteria Environmental Sustainability Index

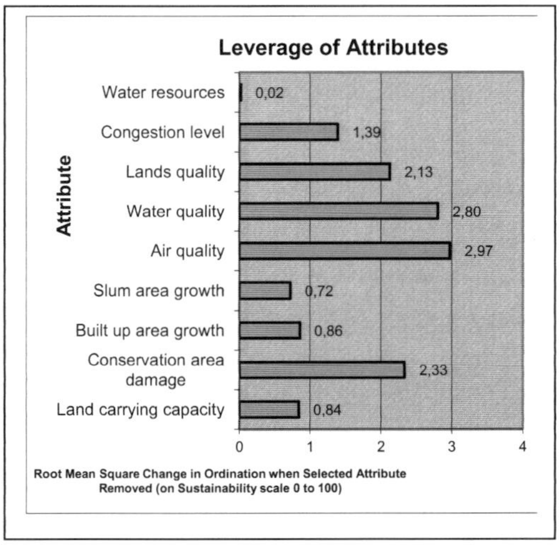

Figure 2. RMS Value of Environmental Criteria

The category was not sustainable due to 7 attributes laid in bad score which were the population growth, the number of poor, artesian/shallow weel by public, catchment area by public, trash processing by public, community behaviour, and safety, security, comfort level. It was also due to 3 attributes laid in moderate score which were HDI, sewage system by public and unemployment rate. The unsustainable status was affected by the 6 key indicators. The RMS of key indicators were: the rate of human development index; the sewage system by public; unemployment rate; trash processing by public; catchment area by public and the making artesian or wells drilled by the public.

The sustainability index value for the economic criteria was 43.88 % which was relatively less sustainable. The category was less sustainable due to all economic attributes laid in moderate score. The less sustainable status was influenced by three key indicators, that the RMS of key indicators were: the rate of investment; level of income per capita, and the local economy growth.

Sustainability index value for technology criteria was 28.32 %. It was classified as less sustainable due to 5 attributes laid in bad score which were sewage system, drinking water system, bicycle lanes/non-motorcycle vehicle, facilities for pedestrians, public transportation. Four (4) attributes laid in moderat score which were drainage systems, solid waste management, green open space systems and road systems. The less sustainable status was influenced by eight key indicators, the RMS of key indicators were: the level of water services; availability of green open space; availability of roads; availability of pedestrian facilities; waste management; availability of municipal sewage system; the availability of bike lanes/non- motorcycle vehicles and the availability of public transport systems.

The sustainability index value of good governance criteria was 44.58 %. It was classified as less sustainable due to 4 attributes laid in bad score which were regulation, inter-sector institution, law enforcement, social political conditions. Five (5) attributes laid in moderate score which were the visionary leadership, spatial planning, budgeting, human resource capacity in goverment, and community participation. Only one attribute laid in good score, it was call center. The less sustainable status was influenced by 5 key indicators, the RMS of key indicators were: law enforcement; call centers; inter-sector institution; leadership, and the local socio-political conditions.

The results of MDS using Rapinfra shows that multicriteria sustainability infrastructure Bandarlampung index value was 38.05 % or less sustainable , as shown in Table 3 and Figure 3.

Sustainable Status			
No.	Criteria	Index value	Sustainable status
1	Environment	42.88	less sustainable
2	Economic	43.88	less sustainable
3	Social	15.80	not sustainable
4	Technology	28.32	less sustainable
5	Governance	44.58	less sustainable

Table 3. Sustainable status of infrastructure

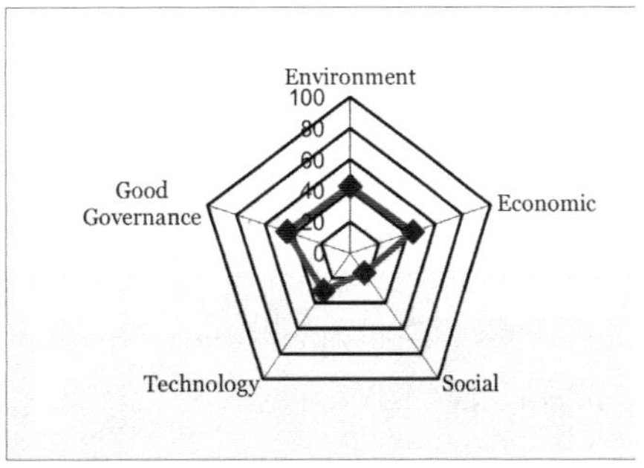

Figure 3 Kite diagram

To determine whether the indicators examined in MDS analysis was quite accurate and can be justified scientifically, this can be seen from the stress and the coefficient of determination (R2). This value was obtained in the MDS analysis using Rapinfra software. The results of the analysis were considered sufficiently accurate and reliable because it has a smaller stress value of 0.25 or 25% and the coefficient of determination (R2) values approaching 1.0 or 100 percent [11]. The analysis showed that all indicators were assessed fairly accurate and accountable. It was shown that the stress value by 14% - 15% and the coefficient of determination (R2) of 0.95%. Stress value indicates the proportion of variance that was not explained by the model. It showed that, the lower the value the better the model MDS stress.

3.2. Priority Indicators Analysis on Sustainable Infrastructure Development

The processes to obtain indicators of sustainable infrastructure development priorities starting from the results of a literature review of 50 indicators, then taken to the FGD, in order to obtain 47 indicators. These indicators is used for the assessment of the various sources which include: stakeholders opinion, communities and planning documents performance. The combined results from the three sources were obtained 27 influential indicators, then this influential indicator was brought to the FGD, in order to obtain 20 indicators were selected to be analyzed in the ANP. ANP analysis results obtained with 8 priority indicator , this stage can be seen in Figure 4.

The community survey showed that 24 indicators of the level of importance according to 5 criteria. Influential indicators for environmental criteria were 5 namely: the level of congestion, water quality, availability of raw water sources, air quality and growth of built up area. There were 5 influential indicators for social criteria, namely: HDI level, level of security and safety, unemployment growth rate, waste management system by community and community behavior. There were 4 influential indicators economic criteria, which include: city minimum wage level, local economy development, the growth of infrastructure budget and economic growth (GDP). Influential indicators for technology criteria were 6 consist of: the availability of drinking water system, waste management system, drainage system, green open space system, wastewater system, and public transport system. There were four influential indicators for governance criteria, namely: visionary leadership, law enforcement, infrastructure planning and infrastructure budget.

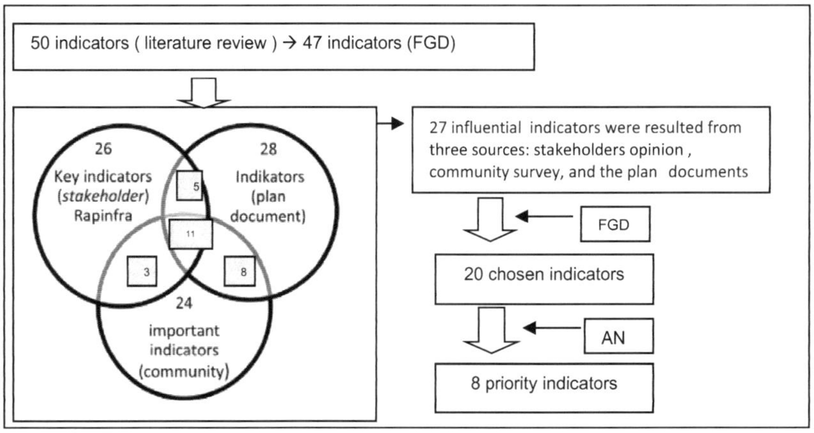

Figure 4. Phases analysis of priority indicator in sustainable infrastucture development

Planning document, which it was the reference of infrastructure development was the Spatial Plan, Sectoral Master Plan and Mid-Term Development Plan (RPJMD). There was also the Mid-Term Infrastructure Plan (RPIJM) space -based and sector and it was currently in the process of preparing the plan. The results showed that the existing infrastructure development plans in the form of a program of activities still does not meet the indicators of sustainable infrastructure development even most plans do not formulate some sort of performance indicators as a measure of development, except RPJMD. Indicators of sustainable infrastructure development on RPJMD were 28 indicators consist of: 6 indicators of environmental criteria, namely: reduced rate of destruction of mountains and hills (conservation area), city slum area, growth of built up area, reduced air pollution, preservation water sources and reduced traffic congestion point. The attributes of social criteria were include 5 indicators, namely: increasing HDI level, reduced the number of poor, security and safety; waste management and unemployment rate. Economic criteria which consist of 5 indicators were: economic growth, city revenue growth, GDP growth rate, the rate of minimum city wage. Criteria technology has 7 indicators namely: growth of number of roads, arranged green open space areas, reduced sedimentation of waterways and drainage, increasing water service, available sewerage installation, available facilities and mass transit traffic. Good governance criteria has 5 indicators, namely: increasing the amount of the approved legislation, capacity building through discipline and education of civil servants, availability of media complaints, budgeting, availability of information planning in accordance with the implementation plan.

The key indicators of MDS previous results (26 indicators) then combined with the results of the community survey (24 indicators) and outcome indicators in the assessment of infrastructure planning documents (28 indicators) to obtain the most influential indicators in the sustainable infrastructure development. Composite indicator made up of indicators that appear at least twice in all three stages of the analysis, in order to obtain 27 indicators (Figure 4).

Figure 4 shown that there were 27 the powerfull indicators in sustainable infrastructure development. The environmental criteria has 6 indicators namely: availability of raw water, air quality, water quality, damage of conservation, growth of built up area development and traffic congestion. The social criteria has 4 indicators consist of: HDI, security and safety, unemployment rate, public participation and citizen behavior. The economic criteria has 4 indicators namely: the rate of investment, income per capita, the rate of the local economy and minimum city wage. The technological criteria has 7 indicators namely: the availability of clean water systems, waste management, green open spaces, road network, drainage system, waste water system, and public transport. The governance criteria has 6 indicators including: visionary leadership, call center, law enforcement and sanctions, infrastructure planning and infrastructure budgets. The 27 powerfull indicators on Figure 4 discussed by experts in the FGD and obtained 20 indicators, and then these indicators will be processed at the stage of ANP (Figure 5).

The calculation result with the super decisions obtained the weight of each indicator, the greater the weight of indicator, the greater the influence of these indicator on sustainable infrastructure development. There were 8 main indicators that the most influential or the priority indicators on sustainable infrastructure development. The priority indicator on economic criteria was local economic growth with weights 0.725. For governance criteria, there were two priority indicators, namely: infrastructure planning with weights 0.475 and infrastructure budget with weights 0.446. On the technological criteria, the priority indicator was the availability of clean water system with weights 0.425. For social criteria there were two priority indicators, namely: community participation with weights 0.418 and people's behavior with weights 0.404. On environmental criteria, there were two priority indicators namely: air quality with a weight of 0.369, and the growth of built up area with weight 0.345.

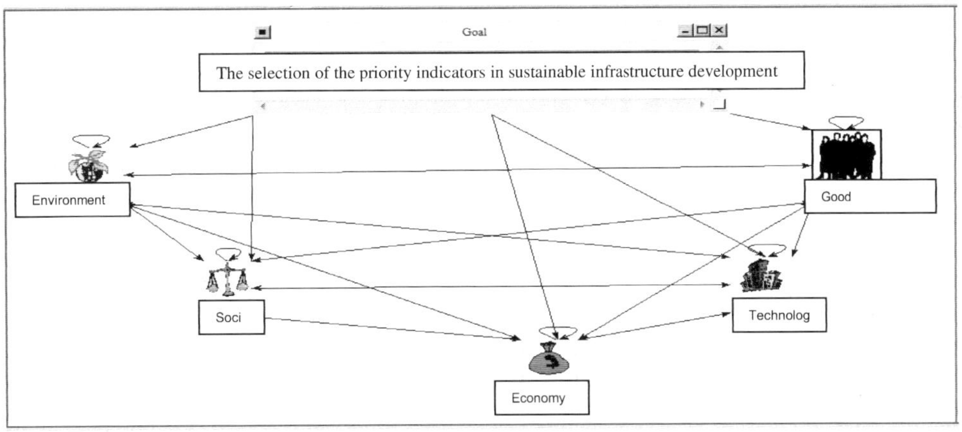

Figure 5. Network link between sustainable infrastructure development criteria in ANP

The results of the analysis of ANP recommend policy directions in the development of sustainable infrastructure ought to consider 8 priority indicators. Policy recommendations in sustainable infrastructure development were the first: the local economic growth that address the needs of micro economic infrastructure such as: provision of space for small enterprise and street vendors in the city. Second: an integrated infrastructure planning between spatial and sectoral development plans. The Indonesian government is currently preparing a program planning development of spatial-based infrastructure to support integrated development through The Medium Term of Infrastructure Investment Program Plan (RPI2JM). This program may be the first step in planning an integrated infrastructure development and sector-based spatial development. This plan can work well if the planning process also involves decision-makers from related sectors. Third: an increase in the infrastructure budget, efficiency and effectiveness of the budget. Fourth: the availability of clean water system which was distributed to all parts of the city, increasing the amount of raw water sources and water management with 5 R (restore, reduce, reuse, recycle, rechargeable). Fifth: increased community participation in the management of city infrastructure, building consensus between the government and the residents of the city as well as the transparency of information. Sixth: city infrastructure management that considers the behavior of (cultural) communities, for example the pattern of movement of people in the use of transport (public transport, bicycle or on foot) and open space utilization patterns. Seventh:

air quality with the increased use of public transport, periodic emissions testing, vehicle age restrictions, environmentally friendly fuel, green industry and waste management without burning. Eighth: the city land use in accordance with the city spatial plan, that requires the provision of 30% open space, minimizing damage to protected areas (mountains, slopes and hills) and the efficient use of space with vertical building development.

3.3. Policy Model of Sustainable Infrastructure Development

Sustainable infrastructure development model of Bandarlampung represents the existence of various factors affecting the availability of the city's infrastructure. This dynamic system model is constructed from three sub- models: the sub-model of social, sub-model of the physical environment , and the sub-model of economic. Sub physical model of the environment consists of sub-models of road system, water resources system, waste sytem and green open space system (Appendix 1 and Appendix 2).

Dynamic simulation models are used to create a scenario of sustainable infrastructure development in Bandarlampung. Simulations carried out for a period of 20 years (2007- 2026). Scenarios developed by simulation intervention on model parameters, namely social (population), economic (medium, small, micro enterprises/MSMEs), physical and environment (clean water, green space, sewage and roads), see table 4. Alternative policy scenario includes four types of scenarios which consisted of 1 scenario without intervention (bussiness as usual), and 3 types of scenarios with interventions namely: pessimistic, moderate and optimistic. The third scenario simulated by changing the parameters of current conditions into better conditions in the future. Optimistic scenario is the scenario that led to the intervention expectations of appropriate conditions of minimum service standards (SPM) which is adapted from the Minister of Settlement and Regional Infrastructure Regulation No. 534/KPTS/M/2001 on Guidelines for Minimum Service Standards, and the Minister of Public Works Regulation No. 01/PRT/2014 on Minimum Service Standards for all public works and Spatial Planning, and the target of achieving the Millennium Development Goals (MDGs) as outlined in Long-term National Development Plan (RPJPN) 2015 -2019.

Table 4. Intervention of Scenario Parameter

Parameter	Existing condition	Pesimistic scenario	Moderate scenario	Optimistic scenario
Social (in-migration growth)	0.013	0.010	0.008	0.005
Local economy				
- Middle enterprise growth (%)	0.10	0.11	0.12	0.13
- Small enterprisesgrowth(%)	0.08	0.09	0.10	0.11
- Micro enterprise growth(%)	0.10	0.11	0.12	0.13
Physical and Environmental				
a. Vehicles				
- Private car age (year)	20	17	14	10
- Motor cycle age (year)	15	12	9	5
b. Area of Public Green Open Space (%)	12.6	15.1	17.5	20
c. Garbage transporting (%)	68	80	90	100
d. Drinking Water Leakage (%)	60	46.7	33.4	20
e. Raw water supply (%)	75	85	95	100
f. Communal WWTP supply (%)	20	40	60	80

Note: WWTP: waste water treatment plan

3.4. Sustainable index of infrastructure development

The result of the simulation scenario on city infrastructure sustainability index value are: the lowest index value of 78.79 and the highest index value 142.08. If the lowest value is assumed to be equivalent to the result of the MDS in the previous section, then the existing condition of the city's infrastructure is also less sustainable. Values reflecting the status of sustainability of infrastructure of the city, the higher the index value, the higher the status of the sustainability of urban infrastructure or the better the quality of the environment as a result of infrastructure development of the city (Table 5 and Table 6).

Table 5. Sustainable index of infrastructure development

Scenario	Sustainable index of infrastructure development						
	Popu-lation	Local economy	Water supply	Waste treatment	Flood index	Traffic jam index	Public green open space
Without intervention	0.761	0.704	0.452	4.092	1.004	3.416	0.405
Pesimistic	0.739	0.813	0.593	2.151	0.972	1.898	0.731
Moderate	0.724	0.927	0.977	2.463	0.952	0.517	0.827
Optimistic	0.703	1.072	1.295	1.118	0.934	0.331	0.870
Weight (AHP)	0.349	0.133	0.136	0.058	0.087	0.152	0.085
Trend (CPI)	negative	positive	positive	negative	negative	negative	positive

Table 6. Infrastructure sustainability index value with CPI

Scenario	Sustainable index of infrastructure development							Total index value	Sustainabl e Status/level
	Popu-lation	Local economy	Water supply	Waste treatment	Flood index	Traffic jam	Public green open space		
Without intervention	92	100	100	27	93	10	100	78.79	less
Pesimistic	95	115	131	52	96	17	180	95.77	fairly
Moderate	97	132	216	45	98	64	204	119.05	good
Optimistic	100	152	287	100	100	100	215	142.08	very good
Weight	0.349	0.133	0.136	0.058	0.087	0.152	0.085		

Note : The weight is the result of assessment using AHP (Analytic Hierarcy Process) ; CPI: *Composite Performance Index*

Model of sustainable urban infrastructure development policies with moderate scenario is chosen the most appropriate scenario to be implemented in the time frame until 2026 with consideration of the availability of the resources available today, include: human resources in the government, budgetary resources, and resources of land. The priority policy for the sustainable development of the city's infrastructure moderate scenario are:

a. Physical and environmental policy field is through an integrated infrastructure management program that includes: (1) the management of water resources, in particular the increase in the volume of raw water, clean water service improvement, reduction of loss/leakage of water, integrated solid waste management, communal waste water treatment plan/WWTP; (2) the provision of mass transit is easily accessible, secure, inexpensive and convenient; and the imposition of restrictions on the age of a motor vehicle; (3) an increase the quantity and quality of public green open space.
b. Social policies that include population control through restrictions on the number of people coming into the city. To support the program it is necessary to measure the change in the form of equitable development of infrastructure to the suburbs and judicial operations.
c. Policy economics is managing the local economy through an increase in the number of medium, small and micro enterprises (MSMEs) by providing access to MSMEs on the provision of urban infrastructure needed such as: cheap transport, accessible water supply and integrated waste management.

4. Conclusion

The sustainable infrastructure development benchmarks were generated in this study which included 5 criterias such as environmental, social, economic, technological and governance, and 47 indicators. The status of Bandarlampung infrastructure sustainability was considered as less sustainable with a score of 38.05 % which means that the availability of the infrastructure was still in good condition. However, it needs to be improved to achieve sustainable infrastructure development.

The ANP analysis recommended that the policy directions in the development of sustainable infrastructure ought to consider 8 key indicators as follows: the local economic growth, infrastructure planning, infrastructure budgets, availability of clean water systems, community participation, people's behavioral, air quality and growth of built up area. The were eight (8) recommendation policies in sustainable infrastructure development. Firstly: the local economic growth that address the needs of micro economic infrastructure such as: provision of space for small enterprise and street vendors in the city. Secondly: an integrated infrastructure planning between spatial and sectoral development plans should consider the indicators of sustainable infrastructure development through The Medium Term of Infrastructure Investment Program Plan (RPI2JM). Thirdly: an increase in the infrastructure budget for more efficiency and effectiveness. Fourthly: the availability of clean water system which was widely distributed throughout the city by increasing the amount of raw water sources and water management with 5 R (restore, reduce, reuse, recycle, rechargable). Fifthly: increased community participation in the management of city infrastructure, building consensus between the government and the residents of the city as well as the transparency of information. Sixthly: city infrastructure management that considers the community behavior, for example the pattern of movement of people in the use of transport (public transport, bicycle or on foot) and open space utilization patterns. Seventhly: air quality with the increased use of public transport, periodical emission testing, vehicle age restrictions, environmental friendly fuel, green industry and waste management without burning. Eighthly: the city land use in accordance with the city's spatial plan, that requires the provision of 30% open space, minimizing damage to protected areas (mountains, slopes and hills) and the efficient use of space with vertical building development.

The system dynamics model of sustainable infrastructure development in the study area is comprised of three sub-models: the sub social model, sub-model of the physical environment, and sub-model of the economy has resulted in the formulation of the sustainability index value of city infrastructure. The index value can be increased in line with the simulation scenarios. The results without the intervention and pessimistic scenario showed a trend is not sustainable with the sustainability level is still relatively low. Results of moderate and optimistic scenarios already showed a better level of sustainability or already ongoing and highly sustainable. Selected scenario is the most moderate scenario might be implemented because of the limitations of existing resources include budgetary resources, human resources in bureaucracy and land resources. Under the moderate scenario, estimated in 2026 at Bandarlampung infrastructure development is still ongoing, but should be anticipated external influences also the city which can lower the level of sustainability.

In order to increase the value of the city's infrastructure and the status of sustainability in order to materialize the scenario chosen, it must be supported by appropriate policies. The findings of priority policy can be used as a basis for intervention in the upgrading of the sustainability of urban infrastructure in the future. So that the policy can be implemented, it is equipped with programs and action change as follows:

a. Policy priorities for sustainable infrastructure development is the policy of physical and environmental fields through an integrated infrastructure management program that includes: (1) the management of water resources in an integrated manner, in particular the increase in the volume of raw water, clean water service improvement, reduction of water leakage and waste management (liquid and solid); (2) the provision of mass transit is easily accessible, secure, inexpensive and comfortable which is supported by the provision of facilities for pedestrians such as sidewalks, pedestrian bridges and public transportation stops, restrictions on the age of a motor vehicle; (3) an increase in the breadth and quality of public availability of green open space through revitalization of existing open space, developing innovative garden (the wall garden, vertical garden and roof garden) and increase the number of parks in the suburbs.
b. Supporting policies to policy priorities are:
    - Social policy that includes population control through restrictions on the number of people coming into the city. To support the program it is necessary to measure the change in the form of equitable development of infrastructure to the suburbs and judicial operations.

Policy management economics is the local economy through an increase in the number of MSMEs by providing access to MSMEs through supporting the provision of basic urban infrastructure in the form of adequate infrastructure.

Recommendation
In order to improve further sustainability infrastructure development of Bandarlampung City, the priority indicators in determining the policy of the city's infrastructure development should be taken into consideration by the local authorities. Model of sustainable infrastructure development policy is a policy concept design that can be adopted in the planning of city infrastructure. The preparation of policy planning and sustainable infrastructure development in urban areas are advised to apply the criteria and indicators generated in this study .

Acknowledgement
The author would like to thank to the colleagues of Bandarlampung Planning and Development Board for supporting data. Thanks should also be conveyed to the Education Ministry of the Republic of Indonesia for supporting funds for this research through BPPS scholarship.

References

[1] Aji A 2000 *Sustainable Green Open Space Management (Case Studi: Kota Bandarlampung)* Dissertation (Bogor : Bogor Agricultural Institute Indonesia)
[2] Andayani S Yowono B E Soekrasno 2012 *J. Civil Engineering* **11** 2 p 148-157
[3] Astuti P Amran T G Herdono 2011 Semarang *J. Industrial Engineering* **VI** 2 p 87-94
[4] Benzerra A Cherrared M Chocat B Cherqui F Zekiouk T 2012 *J.. Environmental Management* **101** p 46-53
[5] Barter P Raad T 2000 *Taking Steps Community Action Guide to People-Centered, Equitable and Sustainable City Transport* Sustainable Transport Action Network for Asia and the Pacific The Sustran Network
[6] Danko C C and Lourenco J M 2007 *A Discussion on Indicators and Criteria for Sustainable City Infrastructure Development* Cost 27 Sustainable Development Policies for Minor Deprived City Communities Evora Workshop
[7] Engel-Yan J Kennedy C Saiz S Pressnail K 2005 *Canadian. J. Civil Engineering* **31** p 45–57
[8] Fauzi A and Anna S 2005 *Modeling of Fisheries and Marine Resources for Policy Analysis* (Jakarta: PT Gramedia Pustaka Utama)
[9] Haghshenas H and Vaziri M *J. Ecological Indicators* **15** p 115–121
[10] Litman T and Burwell D 2006 *Issues J. Global Environmental Issues* **6** 4 p 331-347
[11] Kavanagh P and Pitcher T J 2004 *Implementing Microsoft Excel Software for Rapfish: A Technique for the Rapid Appraisal of Fisheries Status* (Fisheries Centre Research Reports) 12 2 p 75.
[12] Kusbimanto I W 2013 *Models of Sustainable City Transport Infrastructure Development Policies in Metropolitan Mamminasata Southern Sulawesi* Dissertation (Bogor: Graduate School, IPB)
[13] Marimin 2004 *Techniques and Applications: Multi Criteria Decision Making* (Jakarta : Grasindo)
[14] Marvin S and Slater S 1997 *J. Progress in Planning* **48** 4 p 247-313
[15] Mell I C 2009 *Can Green Infrastructure Promote City Sustainability*? (Proceeding of The ICE-Engineering Sustainability) 162 1 p 23-34
[16] Miharja M 2007 *Institutional Approaches in Transportation Planning-Metropolitan Land Use.* Paper presented at the Sustainable Transportation Seminar (Bandung: West Hall Bandung Institute of Technology)
[17] Morrissey J Iyer-Raniga U McLaughlin P and Mills A 2012 *J. Environmental Impact Assessment Review* **33** p 55-65

[18] Muhammadi Aminullah E Soesilo B 2001 *Analysis of Dynamic Systems: Environmental, Social, Economic, Management* (Jakarta: UMJ Press)
[19] Newell J P Seymour M Yee T Renteria J Longcore T Wolch J R. Shishkovsky A 2013 J. Cities **31** p 144-155
[20] Pandit A Jeong H Crittenden JC and Xu M 2011 *An Infrastructure Ecology Approach for City Infrastructure Sustainability and Resiliency* IEEE/PES Power System Conference and Exposition PSCE (Phoenix AZ US)
[21] Pontoh N K and Kustiwan I 2009 *Introductory to City Planning* (Bandung: Bandung Institute of Technology Publisher)
[22] Putri W E F 2013 *Sustainable Analysis of City Green Open Space in Bengkulu City* Jurnal Urip Santoso https://uripsantoso.wordpress.com./tag/ruang-terbuka-hijau. access 25 Mei 2013.
[23] Sahely H Kennedy C A and Adams B J 2005 *Developing sustainability Criteria for city infrastructure system* Canadian. *J. Civil Engineering* **32** p 72-85
[24] Safitri D Chaerul M Sembiring E 2012 *A Multicriteria Decision Model for Organic Solid Waste Management with Analytic Network Process Case Study: Jambi Indonesia* Environmental Engineering Department Graduate School ITB (Bandung: Bandung Institute of Technology)
[25] Saniti D 2012 Alternative Determination of Sustainable Water Supply Systems at Muara Angke *Bandung. J. City and Regional Planning* **23** 3 p 197-208
[26] Setiawati E Notodarmodjo S Soewondo P Efendi A J Otok B W 2013 *Infrastructure Development Strategy for Sustainable Waste Water System by Using SEM Method (Case Study: Setia Budi and Tebet District, South Jakarta J. Procedia Environmental Sciences* **17** p 685-692.
[27] Singh K and Steinberg F 1996 *J. Asia Habitat International* **20** p 1-3.
[28] Steinberg F 2007 *Jakarta: Environmental problems and sustainability J. Habitat International* **31** p 354-365.
[29] Suripin 2004 *Sustainable City Drainage* Systems (Yogyakarta : AndyPress)
[30] Tamin O Z 2007 *Towards Sustainable Transportation System in Bandarlampung City*. Proceeding of Sustainable Transportation Seminar Aula Barat ITB (Bandung: Bandung Institute of Technology).

IOP Conf. Series: Earth and Environmental Science **124** (2018) 012008 doi:10.1088/1755-1315/124/1/012008

APPENDIX-1 : Table Sustainability citeria and incators for different infrastructure system from various studies (2000-2013)

Citeria and indicators	Transportation					Drinking Water			Storm water (Drainage)				Waste water			Solid waste			Green open space		
Environmental criteria:	1	2	3	4	5	6	7	8	9	10	11	12	13	14	15	16	17	18	19	20	21
1. Air pollution (ambient air quality)	X	X	X	X	X	-	-	-	-	-	-	-	-	-	-	X	X	X	X	-	-
2. Emission levels (motor vehicle), GHG	X	-	X	-	X	X	-	-	X	-	-	-	X	-	-	-	-	X	-	-	-
3. Noise pollution (noise level Traffic)	-	X	X	X	X	-	-	-	-	-	-	-	-	-	-	-	-	X	-	-	-
4. Area of green space and network	-	-	-	-	X	-	-	-	-	-	-	-	-	-	-	-	-	-	X	-	-
5. Conversion rate of land (land use)	X	X	-	X	X	X	-	-	X	X	-	-	X	-	-	-	-	X	-	X	-
6. Controlling land use	-	-	-	-	X	-	-	-	-	-	-	-	-	-	-	-	-	-	-	-	-
7. Landscape condition	-	-	-	-	X	-	-	-	-	-	-	-	-	-	-	-	-	-	-	-	X
8. Use of energy (fuel consumption)	X	X	X	X	X	X	-	-	X	-	-	-	X	-	-	-	-	-	-	-	-
9. Use of renewable energy	-	-	-	X	X	-	-	-	-	-	-	-	-	-	-	-	-	-	-	-	-
10. Land degradation	-	-	-	-	X	-	-	-	-	-	-	-	-	-	-	-	-	-	-	-	-
11. The efficiency of movement / mobility	-	X	-	X	-	-	-	-	-	-	-	-	-	-	-	-	-	-	-	-	X
12. Environmental management (3 R) /quality	-	-	-	X	-	-	X	-	-	-	-	-	-	X	X	X	-	X	-	-	-
13. The environmental impact of facilities	-	-	-	X	-	-	-	-	-	-	-	-	-	-	-	-	-	-	-	-	-
14. Protection of wildlife / habitat / biodiversity	-	X	X	X	-	-	-	X	-	-	-	-	-	-	-	-	-	-	X	-	X
15. Water polution/water quality	-	X	X	-	-	X	X	X	X	X	X	X	X	X	X	-	-	-	X	-	-
16. The efficiency of natural resources	-	X	-	-	-	-	X	-	-	-	X	-	-	X	X	-	-	X	-	-	-
17. Material and construction waste	-	-	X	-	-	X	-	-	X	-	-	-	X	-	-	-	-	-	-	-	-
18. Pollution of ground water (ground)/quality	X	-	-	-	-	-	-	-	-	-	-	-	-	-	-	X	X	-	X	-	-
19. Permanent puddle and flood	-	-	-	-	-	X	-	-	X	-	X	-	X	-	-	-	-	-	-	-	-
20. Environment Aesthetic	-	-	-	-	-	-	-	-	-	-	-	-	-	-	-	X	-	-	-	-	-
21. Ecological network (hubs-nodes-corridors)	-	-	-	-	-	-	-	-	-	-	-	-	-	-	-	-	-	-	-	-	X
22. Water resources (air baku)	-	-	-	-	-	-	-	X	-	-	-	-	-	-	-	-	-	-	-	-	-
Economic criteria:	1	2	3	4	5	6	7	8	9	10	11	12	13	14	15	16	17	18	19	20	21
1. The regional Economic Growth	-	-	-	-	X	-	-	-	-	-	-	-	-	-	-	-	-	-	-	-	-
2. Gross Regional Domestic Product	-	X	-	-	X	-	-	-	-	-	-	-	-	-	-	-	-	-	-	-	-
3. Budget implementation infrastructure (OM)	X	X	-	X	X	X	X	X	X	-	X	X	X	X	X	X	X	X	-	-	-
4. The Local Government Revenue	X	-	-	-	X	-	-	-	-	-	-	-	-	-	-	X	-	-	-	-	-
5. Trip fee/service fee	-	-	X	X	X	-	-	-	-	-	-	-	-	-	-	-	-	-	-	-	-
6. Growth centers	-	-	-	-	X	-	-	-	-	-	-	-	-	-	-	-	-	-	-	-	-
7. The growth of land value	-	-	-	-	X	-	-	-	-	-	X	-	-	-	-	-	-	-	-	-	-
8. Dimensions of city area	-	-	-	-	X	-	-	-	-	-	-	-	-	-	-	-	-	-	-	-	-
9. Infrastructure rate/user fee	X	X	X	X	-	X	-	X	X	-	-	-	X	-	-	-	-	-	-	-	-
10. Revenue per capita	-	X	-	-	-	-	-	-	-	-	-	-	-	-	-	-	-	-	-	-	-
11. Supporting industrial/ investment develop.	X	X	-	-	-	-	-	X	-	-	-	-	-	-	-	-	-	-	-	-	-
12. Creation of jobs	-	-	-	-	-	-	-	-	-	-	-	-	-	-	-	-	X	-	X	-	-
13. Absorption of labor	-	-	-	-	-	-	-	-	-	-	-	-	-	-	-	-	X	-	X	-	-
14. Revenue population	-	-	-	-	-	-	-	-	-	-	-	-	-	-	-	X	X	-	-	-	-
15. Local Economic Development (LED)	-	-	-	-	-	-	-	-	-	-	X	-	-	-	-	-	-	X	X	-	X
16. Saving of roads maintenance	-	-	-	-	-	-	-	-	-	-	X	-	-	-	-	-	-	-	-	-	-
Social criteria:	1	2	3	4	5	6	7	8	9	10	11	12	13	14	15	16	17	18	19	20	21
1. Population growth	-	-	-	-	X	-	X	-	-	X	-	-	-	X	-	-	-	-	-	-	-
2. Levels traffic(infrastructure) accidents	-	X	-	-	X	-	-	-	-	-	-	-	-	-	-	-	-	-	-	-	-
3. The Safety level	X	X	X	X	-	-	-	-	-	-	-	-	-	-	-	-	-	-	X	X	-
4. The Security level	-	X	X	X	-	-	-	-	-	-	X	-	-	-	-	-	-	-	X	X	-
5. Behaviour of community as an infras. user	-	X	-	-	X	-	-	X	-	-	-	-	-	-	-	-	-	-	-	-	-
6. The healthy level	X	X	X	-	X	X	X	-	X	-	-	X	X	X	-	X	-	-	X	-	X
7. Growth of private vehicles	-	-	X	-	X	-	-	-	-	-	-	-	-	-	-	-	-	-	-	-	-
8. Education and skills rate	-	-	-	-	X	-	-	X	-	-	-	-	-	-	-	X	-	-	X	-	X
9. Welfare society level	-	-	-	-	X	-	-	-	-	-	-	-	-	-	X	X	-	-	-	-	-
10. Population density	-	X	-	-	X	-	X	-	-	-	-	-	-	X	-	-	-	-	-	-	-
11. Facilities for the disabled	-	X	X	X	X	-	-	-	-	-	-	-	-	-	-	-	-	-	-	-	-
12. Access to public services	X	-	X	X	X	X	-	-	X	-	-	X	X	-	-	-	-	-	-	-	X
13. Satisfaction of road (infrastructure) users	-	-	-	X	X	-	-	-	-	-	-	-	-	-	-	-	-	-	-	-	-
14. Levels Traffic violations	-	-	-	-	X	-	-	-	-	-	-	-	-	-	-	-	-	-	-	-	-
15. Equity / fairness	-	X	X	X	-	-	-	-	-	-	-	-	-	-	-	-	-	-	-	-	-
16. Survival rate	-	-	-	-	-	-	-	-	-	X	-	-	-	-	-	X	-	-	-	-	-
17. Making infiltration wells by community	-	-	-	-	-	-	-	-	-	-	-	-	-	-	-	-	-	-	-	-	-
18. Protection of culture resources/traditional	-	X	X	-	-	-	-	-	-	-	-	-	-	-	X	-	-	-	-	-	-
19. The social interaction and social access	-	-	-	-	-	-	-	X	-	-	-	-	-	-	-	-	-	-	-	X	X
20. Willing to pay	-	-	-	-	-	-	-	X	-	-	-	-	-	-	-	-	-	-	-	-	-
Technology Criteria	1	2	3	4	5	6	7	8	9	10	11	12	13	14	15	16	17	18	19	20	21
1. Capacity of infrastructure (supply)	X	-	-	-	X	X	-	-	X	-	-	-	X	-	-	-	-	-	-	X	-
2. Levels of service (performance)	X	X	X	-	X	X	-	X	X	-	-	-	X	-	-	-	-	-	X	-	-
3. Integration of infrastructure	-	-	-	-	X	-	-	X	-	-	-	-	-	-	-	-	-	-	-	-	-
4. Development of road network	-	X	-	-	X	-	-	-	-	-	-	-	-	-	-	-	-	-	-	-	-
5. Diversification of transportation mode	-	X	-	-	X	-	-	-	-	-	-	-	-	-	-	-	-	-	-	-	-
6. Bike and pedestrian path	-	-	X	-	X	-	-	-	-	-	-	. -	-	-	-	-	-	-	-	-	-
7. Facilities for pedestrian (crossing bridge)	-	-	X	-	X	-	-	-	-	-	-	-	-	-	-	-	-	-	-	-	-
8. Facilities non-motorcycle vehicle	-	-	X	-	-	-	-	-	-	-	-	-	-	-	-	-	-	-	-	-	-
9. Infrastructure technology (design of infra.)	-	-	-	-	-	-	-	X	-	X	-	-	-	-	-	X	-	X	-	-	-
10. The public transport quality	-	-	X	-	X	-	-	-	-	X	-	-	-	-	-	-	-	-	-	-	-
11. Easily obtained/operated of technology	-	-	-	-	-	-	-	-	-	-	-	-	-	X	X	-	-	-	-	-	-
12. Diversification of green open space	-	-	-	-	-	-	-	-	-	-	-	-	-	-	-	-	-	-	-	-	X
13. Leakage	-	-	-	-	-	X	-	-	X	-	-	-	X	-	-	-	-	-	-	-	-
Govemance criteria:	1	2	3	4	5	6	7	8	9	10	11	12	13	14	15	16	17	18	19	20	21
1. Law enforcement and sanctions/control	X	-	-	-	X	X	-	-	X	-	-	-	X	-	X	-	-	-	-	-	-
2. Quality of human resources	-	-	X	-	X	-	-	X	-	-	-	X	-	-	-	-	X	-	-	-	-
3. Community participation	X	X	X	X	-	X	-	X	X	X	X	-	X	-	-	-	X	-	-	X	-
4. Planning	-	X	X	-	-	-	-	-	-	X	-	X	-	-	-	-	-	-	-	-	X
5. Budget development and R & D	X	X	X	-	X	X	-	-	X	X	-	-	X	-	-	-	-	-	-	X	-
6. Regulation / Law	X	-	-	-	-	X	-	X	X	-	X	-	X	-	-	X	X	-	-	X	-
7. Institutional	-	-	-	-	-	-	-	X	-	X	X	-	-	-	-	-	X	-	-	X	-
8. Call Center	-	-	X	-	-	-	-	-	-	-	-	X	-	-	-	-	-	-	-	-	-
9. Cooperation (between regions, private-goverm)	-	-	-	-	-	-	-	X	-	-	-	-	-	-	-	-	X	-	-	-	-
10. Integrated infrastructure institution	-	-	-	-	X	-	-	X	-	-	-	-	-	-	-	-	-	-	-	-	X
11. Conformity with the Spatial planning	-	-	-	-	-	-	-	-	-	-	-	-	-	-	-	-	X	X	-	X	-

Appendix-2: Causal diagram of sustainable infrastructure development model

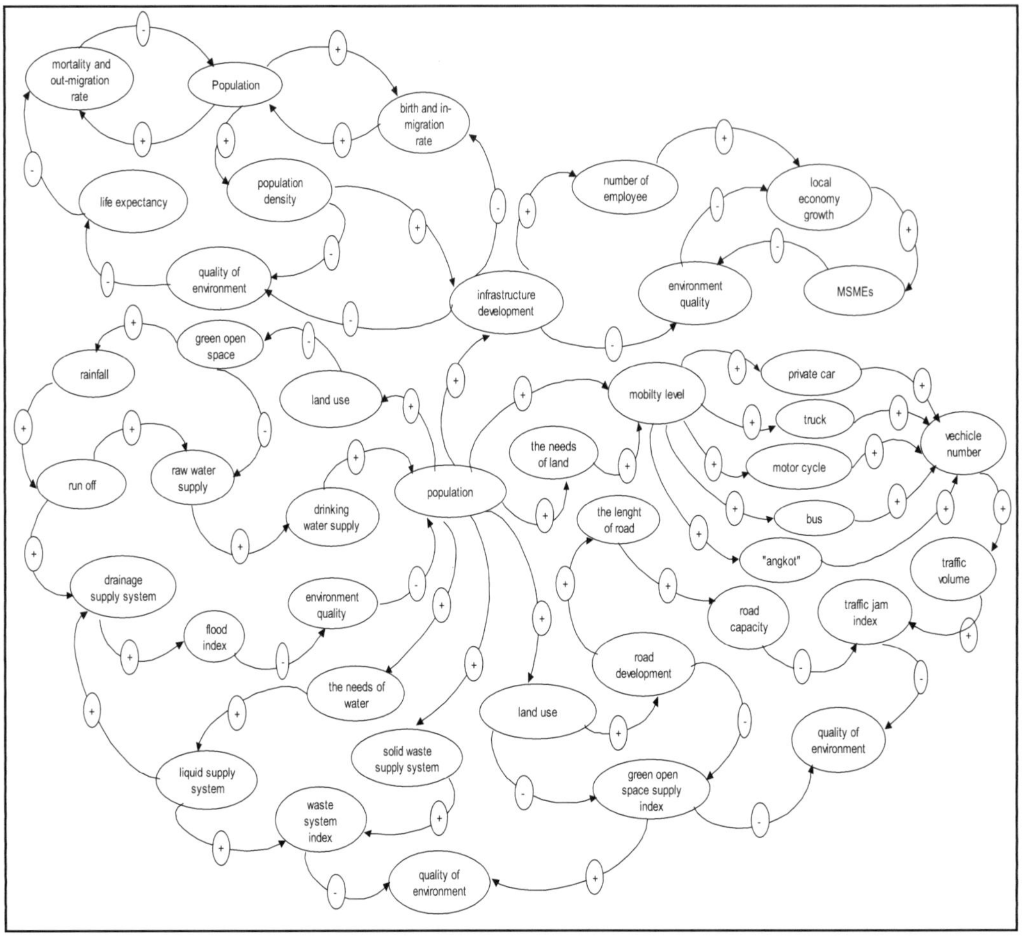

Appendix-3: Stock flow diagram of sustainable infrastructure developm

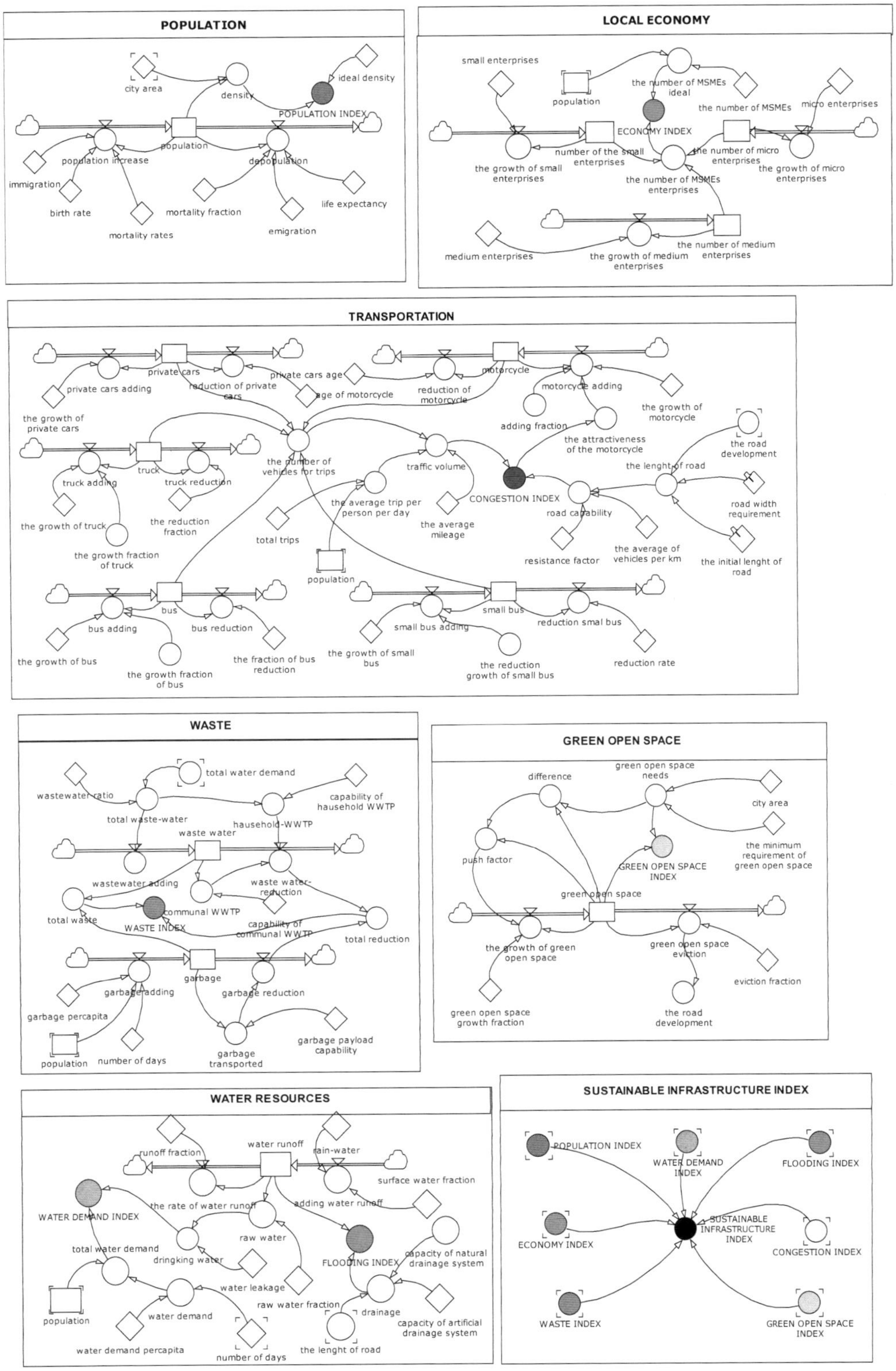

Synthesis of H/Bentonite and Ni/Al_2O_3-bentonite and its application to produce biogasoline from nyamplung seed (*Calophyllum inophillum Linn*) oil by catalytic hydrocracking

A T Marini[1,2], K Wijaya[2], N A Sasongko[1,3]
[1]Department of Energy Security, Indonesia Defense University (UNHAN)
[2]Department of Chemistry, Gadjah Mada University (UGM)
[3] Center for Energy Technology and Chemical Industry, the Agency for Assessment and Application of Technology (BPPT)

Email: asihtrimarini@gmail.com

ABSTRACT. Hydrocracking process of Nyamplung (*Calophyllum inophillum Linn*) seed oil to produce biogasoline using H/bentonite and Ni/Al_2O_3-bentonite that pillared by Al_2O_3 as catalyst had been conducted. Bentonite was activated by acidification using HF 1% and H_2SO_4 0.5 M. Ni metal was impregnated into bentonite with two steps reaction; therewas intercalation with Al_2O_3kegging ion and Ni metal impregnation using $NiCl_2$ metal salt. Catalysts were characterized by infrared spectrophotometer (FTIR), X-ray diffraction (XRD), X-ray fluorescence (XRF), BET, TEM and ammonia adsorption. Hydrocracking reaction was variated by Ni/ Al_2O_3-bentonite and H/bentonite with ratio catalyst/oil 1:100. Biocrude was prepared by extraction by using ethanol 96%. Hydrocracking oil products were further analyzed by GC-MS. The results show that the acidity of bentonite by activation using HF 1% and H_2SO_4 0.5 M has been increased from 62.58 to 64.62 mmol/g. Impregnation process also increased the acidity of bentonite from 62.58 to 64.89 mmol/g. Activation using HF 1% and H_2SO_4 0.5 M, intercalation by Al_2O_3 and impregnation by Ni metal were increasing the crystallinity, surface area, total volume pore and average pore size of bentonite. These techniques were also causeddealumination of bentonite. The hydrocracking process successfully synthesized hydrocarbons with a number of carbon chain between C5-C20 which include bio-gasoline group compounds. Moreover, catalytic processes by H/bentonite and Ni/ Al_2O_3-bentonite also successfully produced 39.83% and 60.37% of biogasoline yields, respectively.

Keyword: activation, bentonite, biogasoline, hydrocracking, "nyamplung",impregnation, intercalation

1. Introduction

In the current of theindustrial era, the problem of energy crisis become an issue that very worrying problem. If decreasing fossil fuel reserves and no reduction in consumption of them, it may cause depletion of the primary energy source fossil source. States more than 90% or 10,000 million tons per year of world energy demand supplied from fossil fuels. If exploitation continues with current figures, it is expected that this energy source will be exhausted in the next half-century [1]. The world's fossil fuel energy reserves, especially petroleum is estimated only 30-50 years later. The most extreme estimates indicate that oil in Indonesia with the current level of consumption is exhausted within 10-15 years [2].

The problem about the depletion of fossil natural resources to make research on alternative energy sources have to be doneespecially for transportation sector that hascontributedto thehigh demand for fossil fuel in Indonesia. Indonesia nowadays is developing renewable energyfor based biodiesel it used for B20 blending and applied for adiesel engine. In the other side transportation sector also dominated by motor cycle usage that consumes gasoline. So not only biodiesel we also have to develop biogasoline to substitute gasoline usage for the transportation sector. Biogasoline is more environmentally friendly, renewable, and the sustainability of its raw material availability is secure. One of the raw materials that can be used is from nyamplung oil. Nyamplung oil obtained from seed extraction from nyamplung plant. Excess nyamplung as araw material is oil from seed has high rendemen, and in its utilization does not compete with the importance of food.Biogasoline can be obtained by hydrocracking process [3,4,5]. The hydrocracking process consists of two stages, combining catalytic cracking and hydrogenation, in which aromatic raw materials with heavy molecules are converted into lighter products in the presence of hydrogen and catalysts. The hydrocracking process can take place using high pressure, high temperature, catalyst, and hydrogen gas, and one of the catalysts that can be used in catalytic hydroxide is modified bentonite [6].

This research is expected to provide information about the use of nyamplung seed oil as one of the alternative energy sources that can produce biogasoline fraction. And know the effect of impregnation and activation treatment on catalyst characteristics used for hydrocracking process

2. Material and Methods

2.1 Literature reviews

Bentonite is one of the types of excavated materials that its availability in Indonesia is quite abundant. Bentonite can be intercalated using polarizing agent one of them is Al_2O_3 which has stable properties so that the application can be used at high temperature (Taslimah, Ratna, and Azmiawati, 2008). The intercalation process is carried out to strengthen the interlayer to improve the physical properties of its aluminum silicate such as crystallinity, basal spacing d001, surface acidity, and thermal stability which is a requirement as a catalyst, catalyst carrier or as an adsorbent. (Figueras, 1988).

Bentonite is one of clay type that has the main content mineral smectite (montmorillonite) with levels of 85-95%. According to (Riyanto, 1992) bentonite may be used as a catalyst support, while modified bentonite may be used as a catalyst. Bentonite is composed of alumina silicate minerals that have a layered crystalline structure. Bentonite has the ability to expand (swellability) due to interlayer space (interlayer) it has, and can accommodate the ions molecules dihydrate with a certain size.

Wahyuningsih (2013) conducted a study on bentonite claiming that the polarization process in bentonite using aluminum oxide (Al_2O_3) increases the physicochemical properties of bentonite, such as observed basal spacing of 0.26 nm, specific surface area 103.93 m^2 / g, and total pore volume 0.06 mL / g but there was a decrease in the mean of pore jores by 2.23 nm.

In the research of Youg-Gou and Ko (2000), the manufacture of alkylated bentonite alumina was carried out by using a polyoxocation $(Al_{13}O_4 (OH)_{24}(H_2O)_{12})^{7+}$ derived from NaOH and $AlCl_3$ solution with a molar ratio of Al / OH = 2. After the calcination of the oxidized alumoxioxide the aluminum becomes Al_2O_3.

The study of Nickel metal expansion in alkyl-pillar bentonite for use as a catalyst in hydrocracking of petroleum fraction has been done by Darmawan (2004). The study suggests that the acidity of bentonite is enhanced by the inclusion of transition metals that have an unfull d orbital.

Table 1. Component of fatty acid constituent ofnyamplung seed oil

Fatty acid component	(%)
Palmic Acid $C_{16:0}$	12,01
Stearate Acid $C_{18:0}$	12,95
Oleic Acid $C_{18:1}$	34,09
Linoleic Acid $C_{18:2}$	38,29
Linolenic Acid $C_{18:3}$	0,3

From the table 1 the highest oil contain of nyamplung seed oil is oleic acid and linoleic acid. Overall nyamplung seed oil consist of the fatty acid chain $C_{16\text{-}18}$.

2.2 Pre-treatment of bentonite

Two kilograms of natural bentonite was soaked in aquadest while stirring with a magnetic stirrer for 24h, then centrifuged to separate the soft and coarse parts both heated to 80-110°C.

2.3 Activation by HF 1%

The soft bentonite was dissolved in 1% HF solution at a 1: 2 ratio and stirred with astirrer for 20min; then the suspension was centrifuged with a 2000 rpm rotational speed for 20 min. The precipitate is separated and neutralized using aquadest to ph 5-6 then dried.

2.4 Bentonit activation by H_2SO_4

50 g of HF-bentonite was dissolvedin 200 ml of 0.5 M H2SO4 solution. The solution was stirred for 1 hour. The precipitate is separated and neutralized using aquadest to ph 5-6 then dried and filtered 250 mesh.

2.5 Development of pillaring agentandintercalation of keggin ion Al_3^+tobentonite

The solution of the pillaring agent was prepared by reacting a 0.4 M $AlCl_3.H_2O$ solution with 0.88M NaOH solution in stirrer for 24 h to homogeny and then suspension of 4g H-bentonite in 200 ml of distilled water and stirred for 20 min. The centrifuges at 2,000rpm for 15 minutes and the precipitate is taken. The precipitate is washed with distilled water until there is no Cl ion. The Cl-Cl content is checked using $AgNO_3$ solution.

2.6 Ni metal impregnation process

2,0244g of $NiCl_2{\cdot}6H_2O$dissolve in 100ml aquadest,then pillared bentonite with 250 mesh size added and stirred using reflux system for 4 hours using 80-90°C of temperature. Then the bentonite is separated and dried using 500°C of temperaturewith theN_2gas flow 20ml/minfor 4hours. Then continued by reduction process using H_2gas flow 20ml/minon temperature 400°C for 4 hours.

2.7 Catalyst acidity test

The catalyst acidity test was performed by NH_3 gas adsorption method.

2.8 Nyamplung seed oil hydrocracking using H/bentonite and Ni/ Al_2O_3 Bentonite

The hydrocracking process conduct by catalyst variation using H-bentonite and Ni/Al_2O_3-bentonite on the oil hydrocracking of nyamplung seed oil. Hydrocracking was used reactor heat 500 ° C for 2 hours with 20 ml/min of H_2 gas flow. The hydrocracking product was analyzed using GC-MS.

3. Results and Discussion

3.1 Acidity test of the catalyst

One of the parameters that affecting the catalitic ability of the catalyst is acidity. Acidity test was performed to determine the acidity change of the catalyst from each treatment. Acidity test was conduct by using NH_3 adsorbtion. The principle of this acidity test is to calculate the change of catalyst weight after NH_4^+ ion is given then cation exchange between H^+ ion and NH_4^+.

Table 2. Result of catalyst acidity measurement

Catayst	Acidity (mol/gram)
Initial Bentonite	62,58
H/bentonite	64,62
Ni/ Al_2O_3-bentonite	64,89

Result of acidity test showed in table 2 that the acidity of bentonite be after treatment is increasing. That is mean that activation of bentonite using acid has been conduct. Then impregnation Ni metal into bentonite also increase the acidity, that it also increase the catalytic ability of bentonite.

3.2 Characterization of catalyst with infrared spectrophotometer (FTIR)

The absorption peak at 1635,64cm^{-1} is the indication of an O-H bending vibration of water molecules. Figure 1 shows the presence of water molecules contained in bentonite minerals.

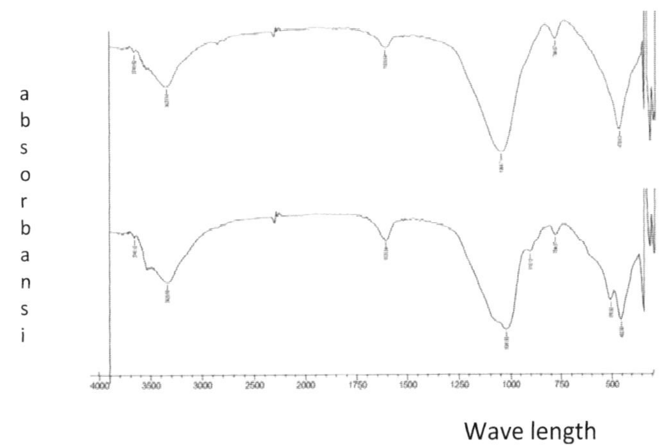

Figure 1. Inffrared spectra (a) H/bentonite dan (b) Ni/ Al_2O_3.bentonite

462,92 cm^{-1}adsorption at H / bentonite and 470.63 cm^{-1} on Ni / Al_2O_3-bentonite shows the bending vibration of Si-O-Si in TO_4 layer, the occurrence of this Si-O-Si vibration wave swing rate indicates the occurrence Dealumination during intercalation and impregnation of Ni metals. Acid treatment does not cause the loss of silica sheet on bentonite. This is good because if the silica layer is damaged, it will decrease the ability of bentonite in its catalytic properties.

3.3 Crystallinity test of catalyst by X-ray diffraction (XRD)

In H / bentonite diffractogram (Figure 2 (c)) seen diffractogram montmorillonite structure shift from the value d = 14,62 Å to d = 13,05 Å this shows during the acidification process with H_2SO_4 0,5M resulted in thecollapse of mineral montmorillonite. The peak of quartz (d = 4,40 Å) and feldspar (d = 4.05 Å) experienced a very significant increase in intensity, possibly due to the collapse of the montmorillonite structure to form quartz and feldspar structures.

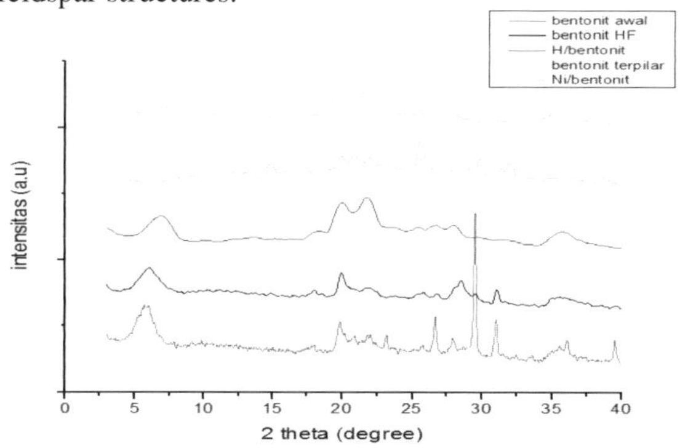

Figure 2. Difraktogram (a) initial bentonite, (b) HF-bentonite, (c) H/bentonite, (d)pillared bentonite (e) Ni/Al_2O_3-bentonite

The Ni/Al_2O_3bentonit diffractogram is showing the montmorillonite peak at d = 14.85 Å where the value is greater than H / bentonite. This shift occurs because of the presence of Ni metals in the bentonite layer, Ni metal with aneutral charge on bentonite will result in a decrease in the number of H_2O molecules being hydrated. After the treatment of Ni metal impregnation, there was still quartz mineral at d = 4.15 Å with relatively high intensity and feldspar mineral at d = 4.02 Å.

3.4 Mineral deposits test on catalysts with X-ray fluorescence (XRF)
Qualitatively, the intercalation process will increase the composition of Al elements in bentonite this is evident from the XRF results show the increase in the composition of Al from 8.2% to 12%. The success of impregnation of Ni metal has been proven by XRF analysis result which shows anincrease of Ni percentage from 0.20 % to 13.3%.

Table 3. Data on XRF analysis result

Element	% in H/bentonite	% in Ni/Al_2O_3-bentonite
Al	8,2	12
Si	47,9	49,7
Ni	0,20	13,3
Others	43,7	25,7

Result of XRF test showed that impregnation using Ni metals has been conduct and it decrease the impurities content of the catalyst.

3.5 Determination of catalyst surface area by BET analysis
The treatment of impregnation of Ni metal on bentonite resulted in a specific surface area of Ni/bentonite of 37.630 m^2 / g, which is less than the surface area of H / bentonite of 79.078 m^2 / g. This shows the successful process of Ni metal embedding in bentonite. The mean pore size of Ni/bentonite is greater than

H / bentonite. The average size of Ni/bentonite pore jores increased due to the presence of Ni metal in bentonite pore reduced the size of pore jores.

Table 4. Analysis Result of BET*analyzer*

Sample	Surface area (m^2/g)	Size pore jores (Å)	Pore volume (10^{-2}cc/g)
H/Bentonite	79.078	88,9709	17,59
Ni/ Al_2O_3-bentonite	37.630	140,563	13,22

3.6 Hydrocracking Product Content Test

The result of this research, hydrocracking of nyamplung seed oil using H / bentonite catalyst produce aproduct with ayield of biogasoline equal to 39,83%. The dominant component of this hydrocracking is a component with a C11 chain of 9.31%, C14 of 9.30% and C16 of 8.22%. Hydrocrackingnyamplung seed oil using Ni/bentonite catalyst produces biogasoline component (C6-12) with ayield of 60,37% product.

4. Conclusion

The hydrocracking process of nyamplung oil with H / bentonite and Ni/bentonite catalysts containing biogasoline fraction. The hydrocracking product contains 39.83% biogasoline fraction on the use of H / bentonite catalyst and 60.37% on the use of Ni / Al_2O_3-bentonite. The influence of bentonite characterization is the presence of Al2O3 pillar on bentonite layer structure increasing interlayer distance on bentonite structure so that Ni metal impregnation process can take place and increase Al amount on bentonite from 8.2% (b / b) to 12% (b / b).

REFERENCES

[1] Jumina, Wijaya, K. dan Budiman, A., 2010, *Dari Yogyakarta untuk Energi Indonesia*, Pusat Studi Energi UGM, Yogyakarta.
[2] Nugroho,A.,2006,*Bioremidiasi Hidrokarbon Minyak Bumi*, Graha Ilmu Universitas Trisakti, Indonesia.
[3] Ainuri, M., 2010, Preparasi Biodiesel dari Minyak Nyamplung (Callophyum Inophylum Linn) Terkatalisis H-Zeolit dan NaOH, *Skripsi,* FMIPA, Universitas Gadjah Mada, Yogyakarta.
[4] Dweck, A. C. dan Meadows, T., 2002, Tamanun (Callophyum Inophylum Linn) the Africa, Asia Polynesia and Pasific Panacea. *International J. Cos Sci* 24:1-8.
[5] Heyne, K., 1987, *Tumbuhan Berguna Indonesia III*, Balai Penelitian dan Pengembangan Kehutanan, Departemen Kehutanan, Jakarta.
[6] Darmawan, A., 2004, Hidrorengkah Fraksi Berat Minyak Bumi Menggunakan Katalis Lempung Terpilar Aluminium Berpengemban Nikel, *JKSA*, 7 (1) 6-10.

The digester modification for biogas production from palm oil mill effluent by Fed-batch

M Aznury[1], J M Amin[1], A Hasan[1], A Harsyah[1]
[1]Department of Chemical Engineering, Politeknik Negeri Sriwijaya, Palembang
Jl. Srijaya Negara Bukit Besar,Palembang 30139, Indonesia

Email : martha_aznury@polsri.ac.id

Abstract. The purpose of this research is to biogas production in the digester modification equipment by Fed-batch of the palm oil mill effluent (POME) to determine the quality of POME after a treatment and the concentration of biogas that is formed every 24 hours within 10 days. The raw materials used are POME from PT Mitra Ogan, Tbk. In the initial stage is sedimentation process in the first digester tank at a flow rate 6 liters/minute and then observing the retention time of 24 hours. POME flowed into the second digester tank for fermentation process with the addition of active microbes seed every 24 hours to produce biogas. After the fermentation process is complete, POME flowed to third digester tank for water treatment stage before being released into the environment. COD content test values obtained after processing are 766, 362 and 350 mg/L, approximately. While the BOD value is 212.75; 125 and 110.9 mg/L, approximately. Biogas production for 10 days fermentation are 10.88% methane, 19.2% oxygen and 75.83% nitrogen, approximately.

Keywords: Digester Modification, POME, Biogas, Fed-Batch

1. Introduction

According to data from the Statistics Book of Palm Oil Commodities published by Directorate General of Plantation (2015) [1], oil palm plantation spread in 32 provinces in Indonesia. Riau Province in the year 2014 with an area of 2.3 million ha and crude palm oil (CPO) production reached 7 million tons is the province that has the largest oil palm plantations and South Sumatra in the 4th largest with an area of 1.11 million ha and production reached 2,85 million tons [1].

CPO production in the palm oil industry produces solid waste and liquid waste. Solid waste consists of empty fruit bunches of palm oil and its waste that has not been utilized properly, as well as shells and fibers that have largely been used as a source of energy by burning it directly. Based on the analysis, 100% production of fresh fruit bunches (FFB) in the palm oil industry will produce 21.5% empty fruit bunches, 22.5% crude palm oil (CPO), and 56% of liquid waste. Liquid waste or palm oil mill effluent (POME) production in Indonesia is estimated around 63 million ton/year [2]. To be able to control the consequences caused by the liquid waste of the palm oil industry, it is necessary to do the processing stages using a biological process involving aerobic bacteria and anaerobes that can be utilized into new products such as biogas that have economic value.

Methane (CH_4) gas is a gas that can cause greenhouse effect causing global warming phenomenon because methane gas has an impact 21 times higher than carbon dioxide (CO_2) gas. Continued reduction of methane gas can play a positive role in addressing global (greenhouse effect) problems that result in global climate change. Wastewater treatment of the palm oil industry is generally done by an open pond system, which consists of several treatment ponds. However, technically these pools are not maintained properly and increasing of methane in the atsmosphere.

For biogas production, POME generated from the three-stage process, namely sedimentation, fermentation, and collection stages. Our research were build of three stages in two shaped such as beam and truncated-pyramid-shaped. According to a final report written by Sari [3], the efficiency comparison between the beam sedimentation tank design and the tank sedimentation tank design for the 6 L/min flow rate is very different. In the design of the beam, the efficiency of the sedimentation process reached 13.9%. While truncated pyramid-shaped efficiency higher reaches 15.6%. So the design of the truncated pyramid-shaped tool has proven that the settling process is faster compared to the beam design tool, so as to minimize the failure in wastewater treatment.

Fermentation system in treatment batch and fed-the batch system. The batch system is fermented by incorporating media and inoculum simultaneously into the bioreactor and the product taking is done at the end of fermentation. Fed-the batch system has been widely applied in various fermentation industries and is relatively easier to use for batch process improvement compared to the batch system.

According to a final report written by Fahlevi [4], the results of the study on a beam and truncated pyramid shape for fermentation tank obtained percent mole of CH_4 for a starter volume of 30% within 4 days. A beam and truncated pyramid-shape had 7.3564% and 6,439%. So the design of the beam-shaped tool has been proven that the fermentation process is better than the design with truncated pyramid-shaped.

Based on this, the authors conducted a study on the process of wastewater treatment of POME by modifying the tool digester with truncated pyramid-shaped for sedimentation and beam-shaped for fermentation. So it is expected with truncated pyramid-shaped sedimentation tank has a faster deposition process while by using the tank fermentation beam-shaped. This design tool is one way more effective and efficient in terms of place, time, and processing costs.

2. Methodology

2.1. Raw material and digester

The raw material in the form of palm oil industry wastewater or Palm Oil Mill Effluent (POME) is done waste treatment by modification of digester through sedimentation and fermentation process. In the design of digester, tools consist of several tools that are important components to support the POME for wastewater treatment. The digester serves to replace an anaerobic pool in conventional systems assisted such as the open pond. The modified digester has truncated pyramid and beam-shape, and there are supporting components such as the centrifugal pump, flowmeter, Tedlar bag, swan neck and vacuum pump. The following is a drawing of the design of the digester tool shown in Figure 1.

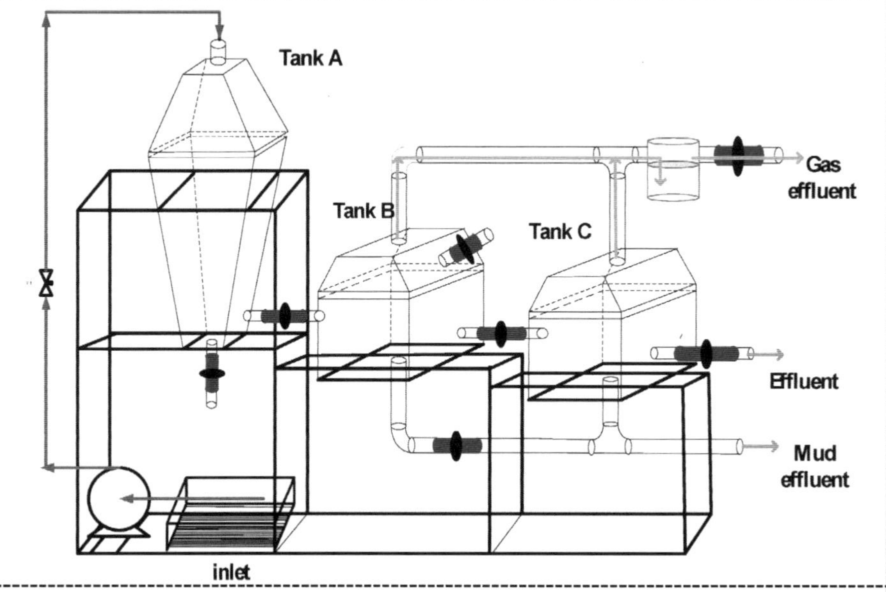

Figure 1. Design of a truncated pyramid-beam Digester

The truncated pyramid digester (tank A) is a container for the initial process of POME. A tank serves as a site for the sedimentation process of the feed. Feed in the form POME which still contains many solid particles or dirt. POME flows from the feed to the tank A. The feed in tank A is carried out by the precipitate to separate the solid particles from the feeder. The collected deposit is separated into a pipeline at the bottom of tank A. This is done so that the dirt/solid particles do not enter in the next tank which can disrupt the fermentation process.

The fermentation process in the digester occurs in the beam shape (tank B). Tank B is a place of degradation of POME by anaerobic fermentation. POME and activity microorganisms (cow manure and POME with 70:30 by fermented at 1 month) in anaerobic processes have the ability to produce biogas. POME rich in organic compounds and carbon, decomposition of organic compounds can produce biogas. After fermented process effluent from tank B is brought to the tank C.

The second beam tank (tank C) as a feeding container for the final processing process of POME. Tank C also serves as a second fermentation process. The bait still has the bacteria that are entangled from the tank B stream and utilized for the fermentation process to re-form methane gas before it is finally discharged into the environment. The precipitates accumulated in the fermentation process are separated through the pipelines at the base of the B and C. The anaerobic conditions can produce residual nutrient-rich wastes such as nitrogen and phosphorus. And the residual liquid waste in the form of fermentation tank can be used for fertilization.

Sedimentation tank supported with one centrifugal pump. The centrifugal pump serves to drain the feed from the feeding point to the A flowmeter tank to adjust the feed flow rate to be flown from the feeding to the tank A. Tedlar bag serves as a gas catcher from the fermentation process and temporary gas reservoir before the methane gas test. The vacuum pump serves to pull the gas from the tank B and the C tank into the Tedlar bag.

2.2. Analytical analysis

The raw material in the form of POME and fermented POME were analyzed by guidelines of the American Public Health Association [5]. Determined of biogas were using a gas chromatograph (GC-8A, Shimadzu).

3. Result and Discussion

3.1. Modified Digester on Truncated pyramid Tank For Sedimentation Process

The sedimentation process for separating water and sludge from the POME in a modified digester in truncated pyramid tank can be determined based on the residence time and flow rate by looking at the results of research conducted by Sari (2015) [3] with using a flow rate of 6 L/m which can achieve the sedimentation efficiency of 15.6% with the observed deposition period that is at 1, 3, 6, 12, and 24 hours with the same feed volume amount 18 liters.

In this research, it is aimed to optimally process of POME by creating a tool that represents the modeling system in waste treatment ponds in the palm oil industry, so it is expected to provide more benefits in treating waste. This research applies the principle of processing done by sedimentation and fermentation process, in the hope that the absence of result from waste treatment is thrown away.

The process that occurs in POME wastewater treatment by using this digester modification is the process of separation between water and sludge with sedimentation (deposition) step by gravity. Where, with the residence time and using a certain flow rate, the raw material in the form of wastewater of palm oil industry will be separated into two layers, the bottom layer is sludge, while the top layer is the wastewater itself. The mud produced will not be thrown away, because the mud from the wastewater treatment can be applied to liquid organic fertilizer.

The sedimentation tool used is a modeling system in waste treatment ponds in the palm oil industry, but different from the model of waste treatment pond commonly used in the form of the beam. The tool used has a truncated pyramid-shaped design based on the results of Sari research (2015) [3], where the sedimentation efficiency obtained from the pyramid design is 15.6% will be separated into two layers of water and mud at the feed rate of 6 L/m. The relationship between the residence time to the volume of water and mud can be seen in Figure 2.

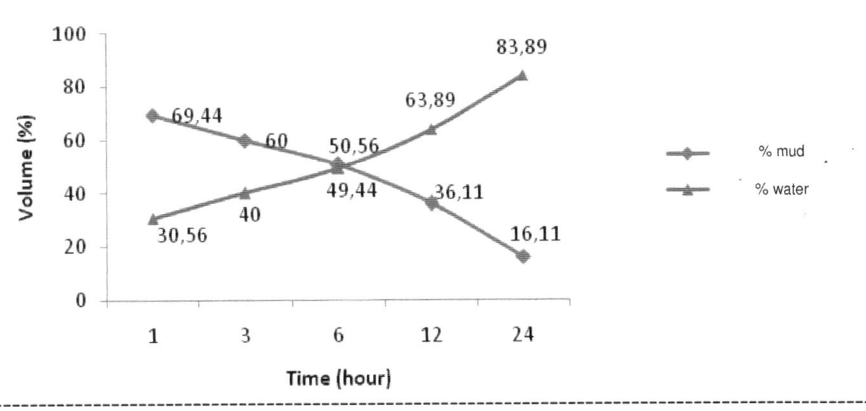

Figure 2. Percentage volume of water and mud at 6 Liter/minute flow rate

Figure 2 showed the percentage of volume water and muddy on truncated pyramid-shaped of the digester (tank A). Percentage mud was 69.44% and 16.11% mud at 1 and 24 hours, approximately. This is because the sedimentation process that occurs in the digester takes place well during the residence time, where the longer the residence time the mud will settle completely and solidify on the base of the digester, causing more water to separate from the mud. According to Fahlevi [4], residence time is directly proportional to the flow rate, so the faster the flow rate it will be faster the time of stay. The longer the residence time, then the volume of water from the truncated pyramid-shaped digester will be constant because the mud has settled completely.

3.2. Analysis Chemical Oxygen Demand (COD) and Biological Oxygen Demand (BOD)

Chemical Oxygen Demand (COD) is used as a measure of the quantity of organic matter in industrial wastewater and has the potential to produce biogas [6]. In the anaerobic renovation process, the decomposed COD is present in the organic material and a final product called methane and new bacterial mass is formed. The measurement of the final COD value in the palm oil industry wastewater after the sedimentation and fermentation process in the digester modification can be seen in Figure 3.

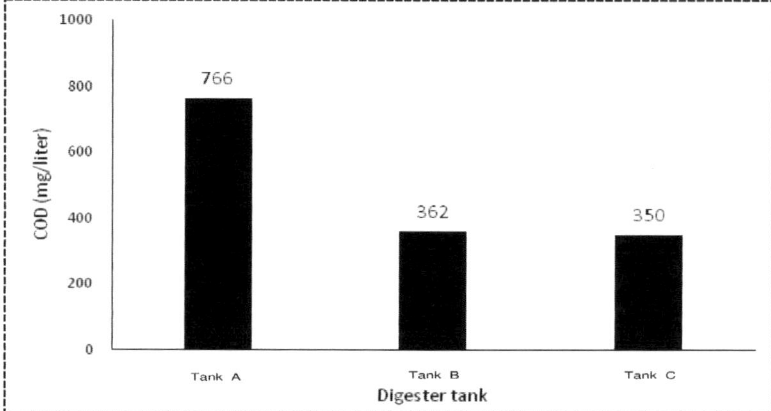

Figure 3. Concentration of COD of POME in Digester Modification

Figure 3 showed the COD values that on the mixture of palm oil and cow dung after the sedimentation process in tank A has decreased the COD value before processing 26352 mg/liter and after processing to 766 mg/liter. The COD value in the fermentation process decreased to 362 mg/liter in the B tank and the final COD value of treatment in the C tank became 350 mg/liter. This indicates that POME waste contains high pollutant load and can be decreased by using anaerobic sedimentation and fermentation processing. The content of organic substances in POME wastes has been degraded by active microbes from cow dung so that the organic material has been shrunk and its COD value has decreased and in the final processing result the COD value has been in accordance with the waste quality standard to be discharged to the environment or the water receiving body. The final result of the processing can also be used as fertilizer in the area of oil palm plantation. The concentration of BOD in POME after fermentation process on digester could be seen in Figure 4.

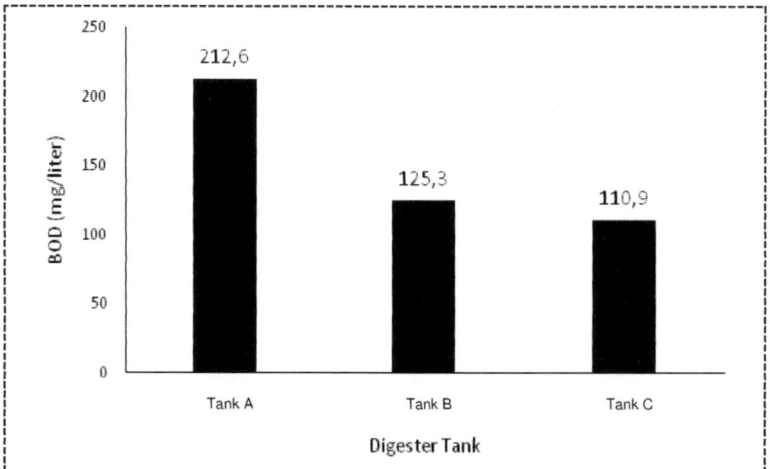

Figure 4. Concentration of BOD of POME in Digester Modification

Figure 4 observed that the decrease in BOD content occurring in POME occurs due to the effect of COD decrease. Where the initial BOD content value before getting the processing on the digester of 1.755 mg/liter. The BOD value after processing has decreased to 110.9 mg/liter in tank C as a result of the final waste treatment product. This is because POME contains a high pollutant load and can go down by using a sedimentation and anaerobic fermentation process. The content of organic substances in POME wastes has been degraded by active microbes that the organic material has been overhauled and the COD and BOD values have declined.

BOD content has decreased, but the BOD content is still high through the standard of waste quality. So waste still needs to be done processing to reduce the BOD content as low as possible based on the regulations made by the Gubernatorial of South Sumatera Number 8 Year 2012 dated February 15, 2012, where the quality standard of waste palm oil industry ready to be released to the receiving river bank or the surrounding environment must have maximum COD 350 mg/liter and BOD maximum 100 mg/liter. The decrease of COD and BOD content occurring after the treatment is due to the fact that some of the organic substances in the waste have been degraded into sludge and some have been degraded into other methane [7].

3.3. Biogas Production Analysis

Wastewater treatment of POME in the modification of digester there are 2 processes consisting of sedimentation and fermentation process. The digester modification on the processing is done by fed-batch system. In the beam-shaped tank (Tank B and C) occurs fermentation stage and addition of substrate. In the fermentation process, the bait decomposed organic compounds by activity microorganisms, the longer the fermentation time the more biogas production increases for 30 days and after that, for 10 days there is a decrease in biogas every day. Therefore, the addition of substrate fed-batch to increase biogas production again. Production of biogas in tank B and tank C is taken using a Tedlar bag to hold gas at the top of the tank beam every 24 hours (1 day) for 10 days.

Biogas is a gas resulting from the final product of digestion or anaerobic degradation of organic materials by anaerobic bacteria in an oxygen-free environment or air [8]. This biogas formation process requires a special installation called a digester, its function is that anaerobic reshuffle can take place well. Therefore, this research treats POME to reduce the harmful chemical compounds. the resulted products from the processing of sludge and biogas (methane) by using digestion modifier. The graph of methane measurement results could be seen in Figure 5.

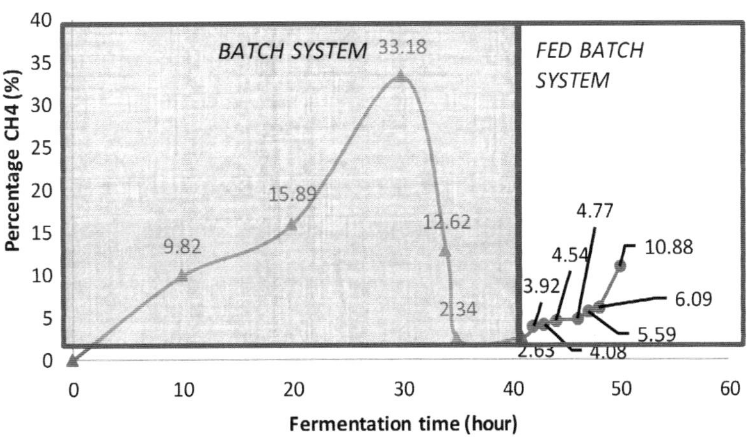

Figure 5. Percentage Methane and fermentation time on fed-batch processing

Figure 5 could be observed the influence of time of fermentation time fed-batch to biogas produced. The fed-batch process there is the addition of new substrate into the tank every day and obtained a single output at the end of incubation for 10 days so that it can be handled in the same way as in batch process. So there is an increase in bacterial growth during fermentation time. On the 1st day it can be seen that the quantity of biogas produced is still low but the longer the fermentation time, the higher the methane gas content in biogas. Fermentation time gives a significant difference to gas production, where fermentation time has a direct effect on biogas production [9].

The content of POME contains high protein and fat and takes several days to experience hydrolysis stage in fermentation to biogas production. At the eco enzymic and anaerobic hydrolysis stage it takes several days to decompose carbohydrates, proteins, and fats into monomers such as amino acids and glucose. Anaerobic bacteria need time to degrade organic substances and certainly can affect the production of biogas.

Fermentation time becomes important in affecting the fermentation process because the process of carbohydrate hydrolysis can be short for several hours. The amount of fat in 39.43 ppm and hydrolysis of protein and fat takes several days until POME does not undergo an effective reshuffle because the methane formation stage has not occurred yet [10]. To form methane needs to experience several stages of hydrolysis, fermentation, acetogenesis, and methanogenesis.

From the graph, it can be concluded that the addition of new substrate in the fed-batch process aims to increase and increase the growth of bacteria so that biogas production that has begun to decline can increase. This is because anaerobic bacteria are added to increase the process of degradation of organic substances and can certainly affect the production of biogas. And the longer the fermentation time the more the content of biogas.

3.4. The Effect of Oxygen Contents on Biogas Production

The concentration of oxygen was high in the digester can disrupt the process of decomposition of organic substances, because organic substances can produce more optimal biogas in oxygen-free [11]. The graph of oxygen and methane gas relations could be seen in Figure 6.

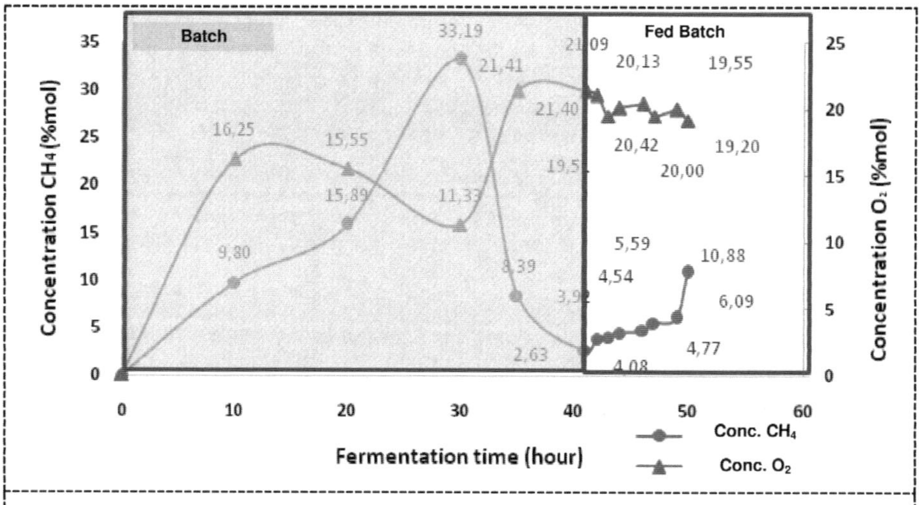

Figure 6. Percentage Methane and oxygen on Fed-batch processing

Figure 6 showed a graph of concentration changes in methane and oxygen gas. That can be observed in biogas production continuously improvement, this is caused by the addition of substrate so that the degradation process is increased to produce biogas. On the contrary, the oxygen content in the digester modification lasts longer then decreases will lead to increased production of biogas. Biogas will form naturally when microorganisms decrease the level of organic matter of POME in the condition of minim oxygen (anaerobic) [12]. From the graph, it can be concluded that increasing the production of biogas associated with oxygen content in the digester. The more oxygen content in the digester decreases, the more the increase in biogas production. Because the process of decomposition or degradation of organic substances carried out by bacteria should be done on conditions of minimal or without oxygen (anaerobic). Biogas is an anaerobic degradation end product gas from organic materials by bacteria in an oxygen-free environment (anaerobic) [8]. The processes that occur in anaerobic treatment are hydrolysis, acidogenic and methanogenesis. Several types of bacteria together gradually degrade organic materials from wastewater [7]. Reaction in the stage of anaerobic is not the influence of oxygen. So the less the oxygen content in the digester the more the increase in biogas production.

3.5. Comparison of biogas production in design of digester

Biogas production is obtained from the fermentation process of POME within the reactor, where the fermentation system used is the fed-batch process. The fed-batch system is a system that adds new media regularly to the closed culture that exists in the reactor so that the volume of culture is increasingly increasing [13]. The fed-batch method of incorporating some sources of nutrients into a bioreactor with a certain volume to be obtained by a product close to the maximum, but the concentration of nutrients source is made constant [11]. Comparison of biogas production results from previous researchers can be seen in Table 1

Table 1. Comparison of Biogas Production Results with Previous Research

Digester shape	% volume starter	System	Fermentation time (day)	% mol CH_4	Reference
Beam	30%	Batch	4	7.356	[14]
Truncated pyramid	30%	Batch	4	6.439	[4]
Beam	30%	Fed-batch	45	25.140	[15]
Modification of Truncated pyramid and beam.	30%	Fed-batch	10	10.882	This research, 2017

From Table 1 compare previous research with this research. Previous research Fahlevi [4] and Saputri [14] produced CH_4 with a maximum of 30% of starter in 4 days fermentation time. The yield of methane was 6.439% mole CH_4 using truncated pyramid digester [4] while according Saputri [14] with beam digester was 7.3564% mole CH_4. So from the data above, the form digester also affect the process of biogas formation. The beam-shaped digester produces more biogas than the truncated pyramid-shaped diets. But the truncated pyramid-shaped digester is more efficient in the sedimentation process than the beam digester.

The biogas production result from Fahlevi [4] and Saputri [14] research is still under biogas production by modification of digester with % mole CH_4 which is 10.8817% mole in fermentation time for 10 days. Production biomethane were increasing in fed-the batch system, was 25.14 mol% at the time the addition of substrate every fifth day for 30 days [15]. The longer the fermentation time the more biogas produced. Biogas production is affected by fermentation time because the time of fermentation is directly related to the amount of time required to pass through the stages of CH_4 formation ie hydrolysis, asedogenogenesis, acetogenesis, and methanogenesis.

4. Conclusion

The modification of digester for biogas production from POME by Fed-batch. The Fed-batch process affects the increase in biogas concentration in the fermentation stage of the digester modification. The determining COD were 26352 mg/L (before treatment) to 766; 362; and 350 mg/L (after treatment). BOD values of water products obtained were 1755 mg/L (before treatment) to 212.7; 125.3 and 110.9 mg/L (after treatment). The results of research in determining the concentration of biogas formed the CH_4 value of the biogas concentration obtained was 2.63% (day 1) to 10.88% (day 10). The value of O_2 from the biogas concentration obtained were: 21.4% (day 1) to 19.2% (day 10). The value of N_2 from biogas concentration obtained is: 78,85% (day 1) to 75,83% (day 10). The less oxygen content (anaerobes) in the modification of the digester the more the increase in biogas production. Future investigations need to focus on scale-up fermentation and feeding strategies with appropriate concentration of POME

References

[1] Statistik Perkebunan Indonesia 2014-2016 Kelapa Sawit, 2015, Direktorat Jenderal Perkebunan, Jakarta , [diunduh tanggal 5 april 2016]
[2] Badan Pusat Statistik. 2014, *Data Luas Lahan Sawit, Produksi, serta Ekspor CPO2009-2015*, [diunduh tanggal 5 april 2016]
[3] Sari, Octa Endang . 2015. *Perbandingan Design Alat Sedimentasi yang Berbentuk Balok dan Limas pada Pengolahan Air Limbah Industri Minyak Kelapa Sawit*. Laporan Akhir. Jurusan Teknik Kimia. Politeknik Negeri Sriwijaya. Palembang
[4] Fahlevi, Muhammad Reza. 2015. *Pengaruh Laju Alir Terhadap Waktu Tinggal Pada Digester Berbentuk Limas Secara Batch Dalam Mengolah Air Limbah Industri Kelapa Sawit Menjadi Biogas*. Laporan Akhir. Jurusan Teknik Kimia. Politeknik Negeri Sriwijaya. Palembang
[5] American Public Health Association (APHA). (1995), *Standard Methods for the examination of Water and Wastewater*, 19th Ed. Washington, D.C.
[7] Deublein D and Steinhauster A. 2008, *Biogas from Waste and Renewabe Resources*, Berlin: WILEY-VCH Verlag GmbH & Co. KGaA, Weinheim.
[8] Siallagan, Nurmay. S.R. 2010. Pengaruh Waktu Tinggal dan Komposisi Bahan Baku pada Proses Fermentasi Limbah Cair Industri Tahu terhadap Biogas. Tesis.Universitas Sumatera Utara. Sumatera Utara
[9] Mujdalipah, Siti, S. Dohong, A. Suryani, and A.Fitria, (2014), Pengaruh Waktu Fermentasi terhadap Produksi Biogas Menggunakan Digester Dua Tahap pada Berbagai Konsentrasi Palm Oil Mill Effluent dan Lumpur Aktif. Skripsi. Program Studi Pendidikan Teknologi Agroindustri, Universitas Pendidikan Indonesia. Bandung.
[10] Hasanah, Hilda. 2011. *Penurunan Beban Pencemar Limbah Cair Kelapa Sawit Melalui Proses Fermentasi Anaerob Menggunakan Digester Anaerob Dua Tahap*. Skripsi. Fakultas Teknologi Pertanian. Institut Pertanian Bogor. Bogor.
[11] Rusmana, Iman. 2008. *Sistem Operasi Fermentasi*, Departemen Biologi FMIPA IPB, Bogor Jawa Barat.
[12] Agustine, Restiyana. 2011. Produksi Biogas dari Palm Oil Mill Effluent (POME) dengan Penambahan Kotoran Sapi Potong sebagai Aktivator. Skripsi. Fakultas Peternakan, Institut Pertanian Bogor. Bogor.
[13] Widjaja T, Hariani N, Gunawan S, and Darmawan, (2010), Teknologi Immobilisasi Sel Ca-Alginat Untuk Memproduksi Etanol Secara Fermentasi Kontinyu dengan *Zymonas mobilis Termutasi*, LaboratoriumTeknologi Biokimia JurusanTeknik Kimia, Fakultas Teknologi Industri Institut Teknologi Sepuluh November, Surabaya.
[14] Saputri, Hardina Apri. 2015. *Pengaruh Persentase Volume Starter dan Waktu Fermentasi pada Pome Terhadap Kuantitas Biogas Menggunakan Digester Balok Sistem Batch.*. Laporan Akhir. Jurusan Teknik Kimia. Politeknik Negeri Sriwijaya. Palembang
[15] Aznury M, Amin J, Hasan A, and Himmatuliza A 2017 *Proc. of the 3rd International Seminar on Fudamental and Application on Chemical Engineering* Vol 1840 (American Institute of Physics) p110012 1-8

Acknowledgments

The authors would like to acknowledge the financial support of **Penelitian Stategis Nasional,** Directorate of Research and Community Service, Ministry of Research, Technology and Higher Education provides funding research project grants with number SPPK:050/SP2H/LT/DRPM /IV/2017 date 25 April 2017. Also, authors acknowledge for PT. Perkebunan Mitra Ogan was a support of POME.

IOP Conf. Series: Earth and Environmental Science 124 (2018) 012011 doi:10.1088/1755-1315/124/1/012011

Theoretical Bound of CRLB for Energy Efficient Technique of RSS-Based Factor Graph Geolocation

Muhammad Reza Kahar Aziz[1], Heriansyah[1], EfaMaydhona Saputra[1],Ardiansyah Musa[2]

[1] Electrical Engineering Department, InstitutTeknologi Sumatera, Lampung Selatan, INDONESIA
[2] Advanced Network Lab., School of Electronics & Computer Engineering, Chonnam National University, Gwangju, SOUTH KOREA
Email: {reza.kahar, heri, maydhona}@el.itera.ac.id, ardi@ejnu.net

Abstract. To support the increase of wireless geolocation development as the key of the technology in the future, this paper proposes theoretical bound derivation, i.e., Cramer Rao lower bound (CRLB) for energy efficient of received signal strength (RSS)-based factor graph wireless geolocation technique. The theoretical bound derivation is crucially important to evaluate whether the energy efficient technique of RSS-based factor graph wireless geolocation is effective as well as to open the opportunity to further innovation of the technique. The CRLB is derived in this paper by using the Fisher information matrix (FIM) of the main formula of the RSS-based factor graph geolocation technique, which is lied on the Jacobian matrix. The simulation result shows that the derived CRLB has the highest accuracy as a bound shown by its lowest root mean squared error (RMSE) curve compared to the RMSE curve of the RSS-based factor graph geolocation technique. Hence, the derived CRLB becomes the lower bound for the efficient technique of RSS-based factor graph wireless geolocation.

1. Introduction

Wireless geolocation technologies have played crucial role in the human life. This is because these technologies are significantly required to enable a lot of location-aware services and applications. Hence, the research in wireless geolocation filed is continuously increasing, mainly since two decades ago, when federal communication committee (FCC) issued the requirement for all wireless equipment to support enhanced 911 (E-911)[1]. However, the wireless geolocation technologieshave been utilized since long time ago before the issuing of the above FCC requirement, i.e., in military application such as the radar to detect the approaching objects, the public service of flights services in managing the aeroplane traffic on the air, as well as the introducing of global position system (GPS)[2]. Currently, the utilization of wireless geolocation technologies are incorporated to vast area of services and applications, i.e., wireless charging, vehicle navigation, precision agriculture, elderly and children monitoring, finding victims of disaster, andillegal radio monitoring[3] -[4].

A lot of techniques for wireless geolocation technologies have also been invented with various approach, i.e., least square[5], methods of moment (MOM)[6], Gauss Newton (GN)[7],[8], and factor graph techniques[9], [10]. We notice that the factor graph based of geolocation is one of the most promising resulting high accuracy of detection and low computation of complexity. This is because the message passing algorithm utilized in factor graph framework efficiently and effectively coordinates the stochastic information obtained from the measurement to estimate the position location of wireless devices. The development of factor graph wireless geolocation technique itself begins since two years after the factor graph framework was firstly introduced in [11]. That is when [12] first utilizes factor graph framework to process the stochastic information of time of arrival (TOA) obtained from the measurement to estimate the position of wireless device. After that time, the development of factor graph for wireless geolocation techniques is significantly increasing where all of those techniques exploit the stochastic information of extracted signal metric of the waveforms, e.g., TOA, time differential of arrival (TDOA), direction of arrival (DOA), received signal strength (RSS), and differential signal strength (DRSS). The basic configuration of wireless geolocation technique with factor graph algorithm is shown in Fig. 1. It should be noted that the monitoring spots are only required by RSS-based technique.

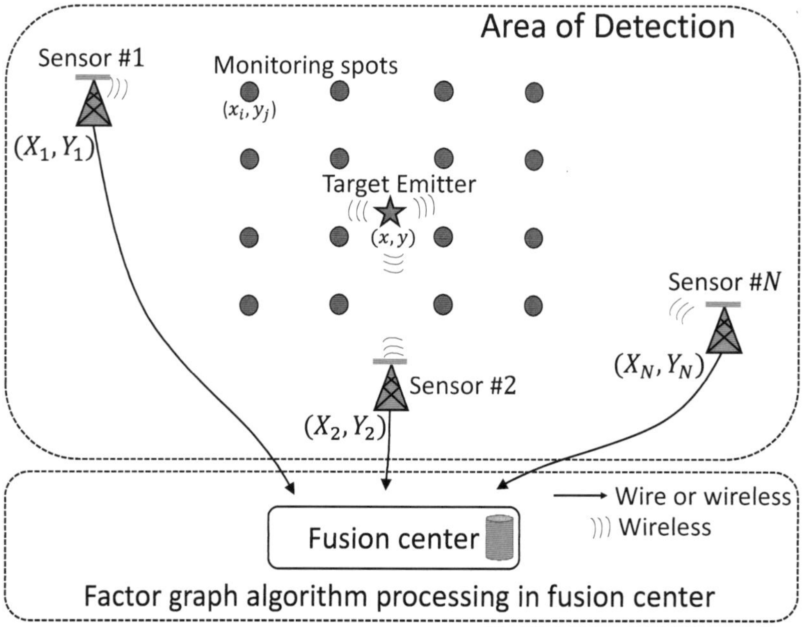

Figure 1. The basic configuration of wireless geolocation with factor graph algorithm.

To ensure the effectiveness and to increase the possibility of the innovation and development of the factor graph based wireless geolocation techniques, the theoretical bound is necessary to be derived. It is well known that the most widely used and easiest bound for any estimation technique is Cramer Rao lower bound (CRLB). Hence, the CRLB for factor graph for TOA[7], TDOA [7], and DOA-based factor graph wireless geolocationtechniques [13] are derived. Furthermore, the introduction of$K$ sample into the CRLB for thosefactor graph techniques is derived in our previous work [14], [15].However, the CRLB for RSS [10] and DRSS-based [16] factor graph wireless geolocation have not been derived yet. Hence, this paper derives the CRLB for RSS-based technique, while the CRLB for DRSS-based technique is left for future work.

2. Model System

Let $\mathbf{x} = [x \ y]^T$ denotes the target position of wireless device in X-Y coordinate system, where$[x \ y]$ and $[\cdot]^T$ are the row vector of the position coordinate and transpose function, respectively. While $\mathbf{X} = [X_i Y_i]^T$ denotes the position of sensor or receiver, where $i, i = \{1, 2, \dots, N\}$,is the sensor index number, while$N$ denotes the total number of sensors. In the RSS-based wireless geolocation system problem, the target position $\mathbf{x}$is estimated with the processing of the RSS value obtained from the measurement. The RSS value itself is the power signal strength received by the sensor or radio receiver. It should be noted that the RSS measurement has been common available in any radio receiver such as smart phone, wireless access point, and other radio receiver devices.

To estimate the target position $\mathbf{x}$, the relationship between the RSS samples ofP_i in each sensor and the target position $\mathbf{x}$ should be available.It is assumed that the P_i follows Gaussian distribussion due to the accumulation effect of many independent factors as mentioned in [10],[9]. This relationship has been derived in [10] for RSS-based factor graph wireless geolocation utilizing the finger print approach[1] expressed as

$$a_{x_i} \cdot x + a_{y_i} \cdot y + a_{p_i} \cdot P_i = c, \quad (1)$$

where a_{x_i}, a_{y_i}, and a_{p_i} are the coefficients of x, y, and P_i, respectively.Furthermore, c denotes a constant with the value of 1 as applied in [10].

It should be noted that the calculation of coefficients of (1) is conducted before the factor graph algorithm is ready for estimating the position of radio emitter. Furthermore, the value of coefficients in (1) for each i-th sensor are obtained by exploiting the measured RSS ofand the coordinate positionof monitoring spots,$P_{i,j}$ and (x_j, y_j), respectively, where$j, j = \{1, 2, \dots, M\}$, is the monitoring spots index number,and$M$is the total number of monitoring spots. In addition, the position and transmit power of monitoring spots are known. Then, the RSS of the signal transmitted by monitoring spots,having the training or pilot signal sent to the sensors, is measured in sufficient long duration to eliminate the error of measured RSS value. Hence, the use of the monitoring spots are for mapping the finger print of RSS value with certain pattern of position by applying least square formula as shown in [10] expressed as

$$\mathbf{A} = (\mathbf{B}^T \cdot \mathbf{B})^{-1} \cdot \mathbf{B}^T \cdot \mathbf{C}, \quad (2)$$

where$\mathbf{A}$is a column vector of a_{x_i}, a_{y_i}, and a_{p_i}, $\mathbf{B}$ is a matrix of x_j, y_j, and $P_{i,j}$. Then, $\mathbf{C}$ is the unit columns vector. Therefore, the equations of (1) having calculated coefficients for each sensorare obtained.

It is shown in [10] and [17]that four monitoring spots with forming square position surrounding the target are the optimal configuration as depicted in Fig. 1.Our work in [17] shows that the estimate accuracies with both area widths of 100×100 m^{2}and 200×200 m^{2}of monitoring spots is sufficient. Furthermore, the algorithm to choose four appropriate monitoring spots located surrounding the target also can be found in our work[18]. The detailed explanation of the RSS-based factor graph technique for wireless geolocation is not discussed in this paper to save the space, however, it can be found in [14] and [10]. Therefore, the next discussion directly goes through to the CRLB derivation for RSS-based factor graph wireless geolocation technique.

It is shown in [2], [7], [8], that the general equations of CRLB is expressed as

$$\text{CRLB} = \sqrt{\text{trace}(\mathbf{F}^{-1}(\mathbf{x}))}, \quad (3)$$

where$\mathbf{F}$ denotes the Fisher information matrix (FIM) as

$$\mathbf{F}(\mathbf{x}) = E\left[\left(\frac{\partial}{\partial \mathbf{x}} \ln p(P_i)\right)^2\right], \quad (4)$$

where$E[\cdot], \frac{\partial}{\partial \mathbf{x}}$,and $p(\cdot)$denote the expectation function, the first derivation respect to $\mathbf{x}$, and the probability density function, respectively.Then, the close form of the CRLB formula with $\mathbf{J}$ being the Jacobian matrix,$\Sigma_P = \sigma_P \mathbf{I}_\mathrm{N}$ being the covariance matrix of RSS sample, $\mathbf{I}$being the identity matrix,

[1]It should be noted that the RSS-based wireless geolocation problem, in general, contains two approach, i.e., (1) finger print and (2) model approach.

Kbeing the measured sample, and $(\cdot)^{-1}$being inverse matrix, is derived from (3) and (4) in our work [14],[15], as

$$\text{CRLB} = \sqrt{\text{trace}\left((\mathbf{J}^{\mathrm{T}}\Sigma_P^{-1}\mathbf{J})K\right)^{-1}}. \tag{5}$$

Then, to obtain the CRLB for RSS-based factor graph geolocation technique, $\mathbf{J}$ should be derived according to (1), where $\mathbf{J} = \frac{\partial P_i}{\partial \mathbf{x}}andP_i = \frac{c}{a_{p_i}} - \frac{a_{x_i}}{a_{p_i}}x - \frac{a_{y_i}}{a_{p_i}}y$.Finally, the Jacobian matrix for RSS-based factor graph wireless geolocation is obtained as

$$\mathbf{J} = \begin{bmatrix} -\frac{a_{x_i}}{a_{p_i}} & -\frac{a_{y_i}}{a_{p_i}} \end{bmatrix}. \tag{6}$$

3. Numerical Results

The computer simulation is conducted to validate the derivation of the CRLB utilized as the lower bound of the RSS-based factor graph wireless geolocation technique. Then, 10,000 times of trials are repeated for smoothing to the curves of the accuracy of the CRLB and the RSS-based factor graph technique.Furthermore, the accuracy itself is defined with the performance of root mean square error (RMSE).Note that three sensors are applied with located at $(100, 0), (1100, 0)$, and $(600, -1000)$m. Then, the target locations are randomly chosen in the centre of area width of $600 \times 600\ \mathrm{m}^2$. In addition, two scenarios of monitoring spots area width are $100 \times\ 100\ \mathrm{m}^2$and $200 \times 200\ \mathrm{m}^2$. Finally, the signal to noise ratio (SNR) values for testing the algorithm and formula are 0 to 45 dB with the step of 5 dB.

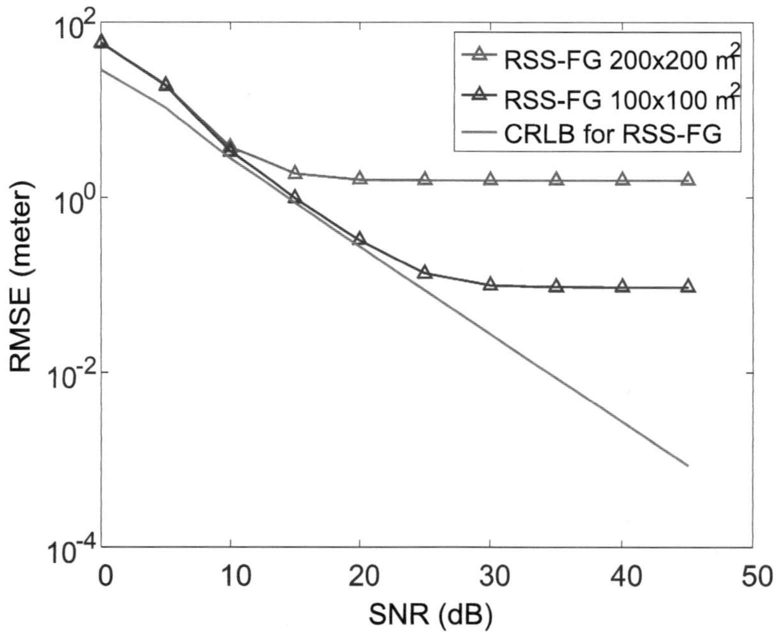

Figure 2. The RMSE of RSS-based factor graph and its CRLB. Note that RSS-FG in the figure stands for RSS-based factor graph technique.

As mentioned above, the detailed of RSS-based factor graph wireless geolocation algorithm can be found in [14], [10], hence we do not present in this paper. Furthermore, 100 iterations are conducted in each trial of the factor graph algorithm simulation. It is assumed that the RSS signal is suffered from path-loss, where in the simulation uses exponential path-loss model with $d_0 = 100$ m being distance reference, $n_p = 3$ being the exponential value, and $f_c = 1$ GHz being the carrier frequency.

Fig. 2 shows that in term of the accuracy of RMSE, the derived CRLB has been the lower bound for both of the RSS-based factor graph wireless geolocation techniques with 100×100 m^2 and 200×200 m^2 of monitoring spots width area. It is also shown that the larger area of monitoring spots the lower of the accuracy is obtained. Furthermore, the accuracy of the factor graph algorithm becomes stagnant above 15 dB and 30 dB for monitoring spots width area of 200×200 m^2 and 100×100 m^2, respectively, while the higher SNR the higher accuracy of the CRLB with linear curve shape.

It is also shown in Fig. 3 that the CRLB curve accuracy located at the lowest RMSE of each iteration which means the CRLB is the most accurate served as the lower bound. It should be noted that the lower of RMSE, the higher accuracy of the estimate. Furthermore, Fig. 3 depicts that the iteration of the algorithm of factor graph converges about at 16-th iteration where the RSS-based factor graph with lower area width of monitoring spots outperforms the wider one.In addition, it is shown in [10] that the computational complexity of the RSS-based factor graph wireless geolocation technique is linearly proportional to N. Hence, the CRLB becomes the lower bound of the energy efficient technique of RSS-based factor graph wireless geolocation.

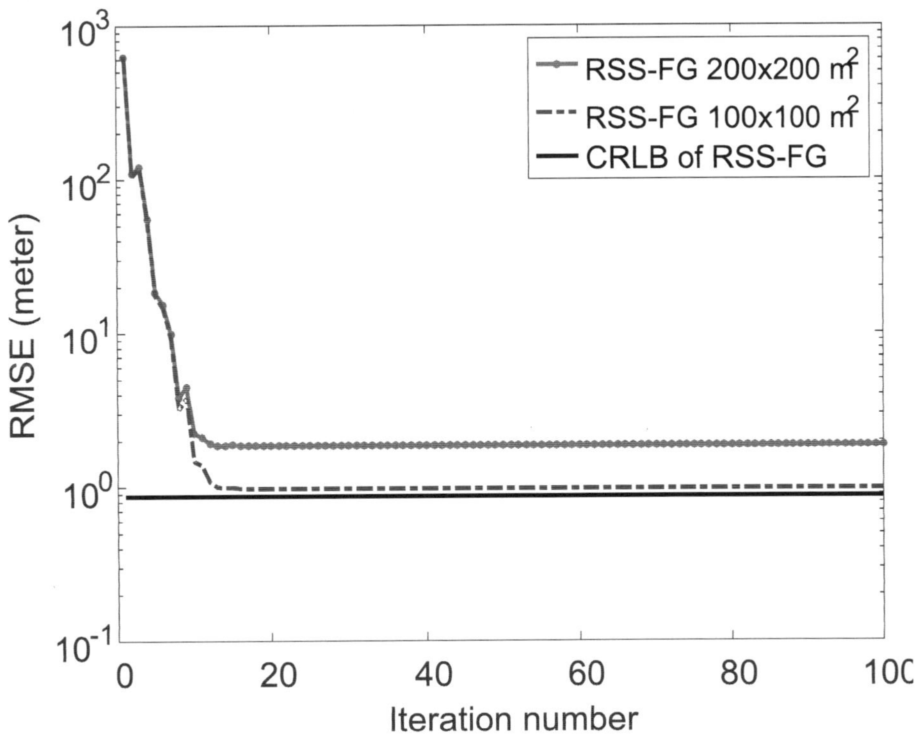

Figure 3. The RMSE of RSS FG and CRLB for RSS FG. Note that FG stands for factor graph.

4. Conclusion

The derivation of CRLB for RSS-based factor graph wireless geolocation has been presented in this paper. It should be noted that the key of the derivation is lied on Jacobian matrix inside the CRLB formula. Hence, the Jacobian matrix is derived from the expression of the relationship between RSS with the coordinate position utilized by the low complexity of RSS-based factor graph wireless geolocation algorithm. Finally, the simulation results shows that the derived CRLB has the highest accuracy compared to the RSS-based factor graph wireless geolocation technique. It means that the derived CRLB is the theoretical lower bound of the energy efficient of RSS-based factor graph wireless geolocation technique.

Acknowledgments
This research is supported by Hibah Mandiri(Grant) of InstitutTeknologi Sumatera in 2017.

References

[1] Caff ery J J and Stuber G L 1998 Overview of Radiolocation in CDMA Cellular Systems*IEEE Communications Magazine***36** 38 – 45
[2] Gentile C, Alsindi N, Raulefs R and Teolis C 2013*Geolocation Techniques, Principles and Applications* (New York: Springer)
[3] Staff ord J V 2005*Precision Agriculture '05* (Wageningen Academic Publishers)
[4] Ackerman E 2016 Wireless Power Takes Charge*IEEE Spectrum*
[5] Kulakowski P, Vales-Alonsob J, Egea-Lopez E, Ludwin W and Garca-Haro J 2010 Angle-of-arrival localization based on antenna arrays for wireless sensor networks*Computers and Electrical Engineering***36** 1181 – 86
[6] Qiao T and Liu H 2013 An improved method of moments estimator for TOA based localization IEEE Communications Letters171321 – 24
[7] MensingC and Plass S 2007 Positioning Based on Factor Graphs*EURASIP Journal on Advances in Signal Processing*
[8] Kay S M 1993 *Fundamentals of Statistical Signal Processing: Estimation Theory* (Prentice Hall)
[9] Chen J –C, Wang Y –C, Maa C –Sand Chen J –T 2006 Network Side Mobile Position Location Using Factor Graphs*IEEE Trans. on Wireless Comm.***5**2696 – 2704
[10] Huang C –T, Wu C –H, Lee Y –N and Chen J –T 2009 A Novel Indoor RSS-Based Position Location Algorithm Using Factor Graphs *IEEE Trans. on Wireless Comm.***8**3050 – 58
[11] Kschischang F R, Frey B J and Loeliger H –A 2001 Factor Graphs and the Sum-Product Algorithm*IEEE Trans. on Inf. Theory***47**498 – 519
[12] Chen J –C, Maa C –S and Chen J –T 2003 Factor Graphs for Mobile Position Location*in Proc. IEEE International Conference on Acoustics, Speech, and Signal Processing (ICASSP)*
[13] Omidali B and Shirazi S A –A B 2009 Performance Improvement AOA Positioning Using Two-Step Plan Based on Factor Graphs and the Gauss-Newton Method*in Proc. 14-th International CSI Computer Conference (CSICC)*
[14] Aziz M R K 2016 *Factor Graph-Based Geolocation Techniques for Position Detection Of Unknown Radio Wave Emitters, Doctoral Dissertation, Japan Advanced Institute of Science and Technology (JAIST)*
[15] Aziz M R K, Anwar K and Matsumoto T 2016 A New DOA-based Factor Graph Geolocation Technique for Detection of Unknown Radio Wave Emitter Position Using The First-Order Taylor Series Approximation*EURASIP Journal on Wireless Communications and Networking (JWCN)*
[16] Aziz M R K, Anwar K and Matsumoto T 2016 DRSS-based Factor Graph Geolocation Technique for Position Detection of Unknown Radio Emitter*in Proc. International Workshop on Advanced PHY and MAC Layer Design for 5G Mobile Networks and Internet of Things in conjunction with European Wireless2016*
[17] Aziz M R K, Anwar K andMatsumoto T 2015 Monitoring spot configuration of RSS-based factor graph geolocation technique in outdoor WSN environment*in Proc. IEICE General Conference 2015*
[18] Aziz M R K, Lim Y and Matsumoto T 2016 A New RSS-based Wireless Geolocation Technique Utilizing Joint Voronoi and Factor Graph*International Journal of Simulation: Systems, Science and Technology***17**9.1 – 9

IOP Conf. Series: Earth and Environmental Science **124** (2018) 012012 doi:10.1088/1755-1315/124/1/012012

Factors Relating Infrastructure Provision by Developer in Formal Housing

H T Putri[1,2], S Maryati[1], and A N S Humaira[1]

[1]Research Cluster for Regional and City Infrastructure System, School of Architecture, Planning, and Policy Development, Institute of Technology Bandung, Jalan Ganeca 10 Bandung, Indonesia.

[2]hsntiara@gmail.com

Abstract. In big cities, housing developer has significant role in infrastructure provision. Nevertheless in some cases developers have not fulfilled their role to complete the housing with infrastructures needed. The objective of this study is to explore the characteristics and the related factors of infrastructure provisioning in formal housing developed by developer using the quantitative and association method analysis. Infrastructures are focused on clean water, sewage, drainage, and solid waste system. This study used Parongpong District, West Bandung Regency as case study where the need of infrastructure is not fulfilled. Based on the analysis, can be concluded that there are some variation in infrastructure provisioning and the factor related the condition is the level of income of house owner target.

1. Introduction

Housing needs is increasing every year. Especially in urban areas, the needs is increasing along the population and urbanization growth. The rapid development of urban areas will give a trickling down effect so that the people needs which can not be accommodated in urban areas will be scattered in the surrounding area, including housing needs. This phenomenon also happened in Bandung, where the rate of population growth in the surrounding area is increasing and activity centers are just beginning to develop. West Bandung Regency became one of the areas that received the impact of Bandung development. It is also confirmed by Vice Regent West Bandung, Yayat Soemitra [1] if the population growth rate of West Bandung regency in 2015 is increased 2.3% and nearly 40% of the population comes from outside the area. As a result, the housing needs is shifting. Meeting its need, housing developer start to develop some new housing include in Parongpong District.

Parongpong is one of the districts in West Bandung Regency which has great appeal to be used as a residential area. In order to the physical environmental conditions that are still relatively cool and sightly, District Parongpong also have a good access to activity center in Bandung. Parongpong has 4,012 hectares area with an average elevation of 700 meters above sea level which consists of 7 village and 30 sub-vilage. The population in 2014 according to the Central Bureau of Statistic[1] reached 102.876 inhabitants and the population density ranges from 22.79 inhabitants/ha.

[1] The Central Bureau of Statistics is a Non-Government Organization that is directly responsible to the President to take care of the population census

The high of housing needs is seen by developers as a great potential businesses. Housing development began to bloom on low price land, but still have good access to urban areas. Based on the Parongpong Profile in 2009 [2] there were 33 housing scattered throughout the village. In addition, according to the Real Estate Indonesia[2] for West Java Province [3] over the last two years, there are 9 filing new housing projects located in the Parongpong, including Bumi Asri Cihanjuang, Green Valley Cihanjuang, Charm City View, Setra Regency Townhouses, and etc.

In developing housing, developer often face a number of obstacles that would make them to do some efforts to minimize production costs, such as Woodson [4] stated: 1) Select the site locations that are not easily accessible and far from settlement services. Away from work, so the living cost in this housing could be higher; 2) Choose the site with lower feasibility occupants, such as flooding and landslide-prone; 3) Reduce the quality of housing infrastructure and utilities also settlement services such as roads easily damaged road, limited clean water, improper sanitation, none waste management facilities, and etc; 4) Reduce the quality of construction by using substandard building materials; and 5) Repeal the security rights guarantee where land certificates were never resolved.

According Winarso [5], the developer has financial ability to build their own infrastructure. Winarso also explained that in order to reduce the cost of its production, developers choose the lower price land despite not having access to basic infrastructure services then built housing that will eventually be equipped with required facilities. The quality of infrastructure provided by the developer also generally have better quality than what is supplied by the government. Thus, in the context of the wider region, Firman [6] mentions that this can lead to social segregation in residential areas to the neighborhood settlements surrounding communities.

Housing developer has significant role in housing and infrastructure provision, especially in big cities. Nevertheless in some cases developers have not fulfilled their role. In this case the need of infrastructure of the community cannot be fulfilled. Then the objective of this study is to explore the condition of infrastructure in formal housing developed by developer and the factors influencing the condition.

2. Methods

The methods described in this section include methods in data collection and analysis.

2.1. Data Collection

The type of data required in the study are primary and secondary data. Primary data is data obtained from observations directly related to the object of research studies that set. Primary data was collected through observation, interviews, and questionnaires to developers and other stakeholders. Based on the steps being taken in research, data collection is done starting from literature, institutional survey, and a survey in the form of direct observation and questionnaires.

Observations are divided into two stages. The first stage is pre survey where observation was conducted to determine the distribution of housing in the district Parongpong which will be the object of research. At this stage of pre-survey, a total number of 55 housing in Sub Parongpong but only 49 housing can be made the object of research to answer hypotheses and research questions have been formulated. The object of research is described in Table 1 and the spatial distribution is showed in Figure 1. Furthermore, at this stage of the survey, observation was conducted to determine a general overview of basic infrastructure that exist in each housing.

Table 1. List of formal housing developed by developer in Parongpong. Type of housing classification: A = low income housing; B = low-middle income housing; C = middle-high income housing; and D = high income housing. Cluster Bella Cassa, Royal Duta Townhouse, Sariwangi Regency II, Setra Regency Townhouse, The Green Garden Massuraya Village, and The Philosopher Village is not observed.

---

[2] REI is the formal housing association in Indonesia

IOP Conf. Series: Earth and Environmental Science **124** (2018) 012012 doi:10.1088/1755-1315/124/1/012012

No	Name Of Housing	Village	Year of Construction	Total Unit	Class
1	Alam Sariwangi	Sariwangi	2009-2010	21-40	B
2	Arthabahana	Cihanjuang	before 2000	151-200	B
3	Bukit Sariwangi	Sariwangi	2005-2006	101-125	A
4	Bukit Taman Asri Residence	Sariwangi	2010-2011	21-40	B
5	Bumi Asri Cihanjuang	Cihanjuang	2010-2011	101-125	B
6	Bumi Sariwangi 1	Sariwangi	2000-2004	201-500	C
7	Bumi Sariwangi 2	Sariwangi	2005-2006	21-40	A
8	Cibaligo Permai	Cihanjuang	before 2000	151-200	B
9	Cluster Bella Cassa	Ciwaruga	2016	Below 21	B
10	Duta Asri Residence	Cihanjuang	2014-2015	21-40	B
11	Graha Sariwangi Residence	Sariwangi	2009-2010	21-40	B
12	Green Hanjuang	Cihanjuang	2010-2011	21-40	B
13	Greenland Ciwaruga	Ciwaruga	2012-2013	41-60	B
14	Griya Amanda 1	Sariwangi	2005-2006	21-40	A
15	Griya Amanda 3	Ciwaruga	2005-2006	21-40	A
16	Griya Cihanjuang	Cihanjuang	2012-2013	21-40	A
17	Harmony Cihanjuang	Cihanjuang	2012-2013	61-80	C
18	Insulinde Indah	Cihanjuang Rahayu	2000-2004	151-200	A
19	Komplek Pasir Kemiri	Cigugur Girang	2007-2008	61-80	A
20	Kompleks Graha Lista	Ciwaruga	2007-2008	21-40	A
21	Kompleks Lembah Sariwangi	Sariwangi	2000-2004	151-200	B
22	Lembah Hijau Cihanjuang	Cihanjuang	2012-2013	61-80	B
23	Lembah Permai Hanjuang	Cihanjuang	before 2000	126-150	B
24	Maharani Village	Cigugur Girang	2014-2015	21-40	B
25	Nirwana Regency	Sariwangi	2000-2004	81-100	B
26	Parigi Indah Kavling DPRD	Ciwaruga	2005-2006	151-200	C
27	Perumahan Katumiri	Cihanjuang Rahayu	2000-2004	201-500	D
28	Pondok Bunga Sariwangi 1	Sariwangi	2007-2008	41-60	A
29	Pondok Bunga Sariwangi 2	Sariwangi	2007-2008	21-40	A
30	Pondok Hijau Indah	Ciwaruga	before 2000	201-500	C
31	Puncak Sariwangi Asri	Sariwangi	2005-2006	21-40	A
32	Puri Budi Asri	Cihanjuang	before 2000	126-150	B
33	Royal Duta Townhouse	Ciwaruga	2016	-	B
34	Royal View Residence	Ciwaruga	before 2000	41-60	C
35	Sariwangi City View	Sariwangi	2009-2010	151-200	B
36	Sariwangi Estate	Sariwangi	2010-2011	21-40	B

IOP Conf. Series: Earth and Environmental Science **124** (2018) 012012 doi:10.1088/1755-1315/124/1/012012

No	Name Of Housing	Village	Year of Construction	Total Unit	Class
37	Sariwangi Highland	Sariwangi	2014-2015	below 21	C
38	Sariwangi Regency II	Sariwangi	2016	-	B
39	Selaras Alam Rumah Inspirasi	Sariwangi	2005-2006	61-80	C
40	Selaras Cihanjuang	Cihanjuang	2012-2013	21-40	B
41	Serra Valley	Ciwaruga	2010-2011	101-125	C
42	Setiabudi Regensi	Ciwaruga	before 2000	501-1000	D
43	Setra Regency	Ciwaruga	2012-2013	21-40	B
44	Setra Regency Townhouse	Ciwaruga	2016	21-40	B
45	SSP Sariwangi Cozy Living	Sariwangi	2009-2010	below 21	B
46	SSP Sariwangi Perumahan Berteknologi	Sariwangi	2009-2010	81-100	C
47	Taman Cihanjuang 1	Cihanjuang	2000-2004	81-100	B
48	Taman Cihanjuang 2	Cihanjuang	2014-2015	21-40	B
49	The Garden Sariwangi	Sariwangi	2007-2008	21-40	A
50	The Green Garden Massuraya Village	Cigugur Girang	2016	-	B
51	The Green Sariwangi	Sariwangi	2007-2008	41-60	A
52	The Philosopher Village	Cigugur Girang	2016	-	B
53	The Sariwangi Village	Sariwangi	2012-2013	41-60	B
54	Trinity	Cihideung	before 2000	61-80	D
55	Villa Istana Bunga	Karyawangi	before 2000	201-500	D

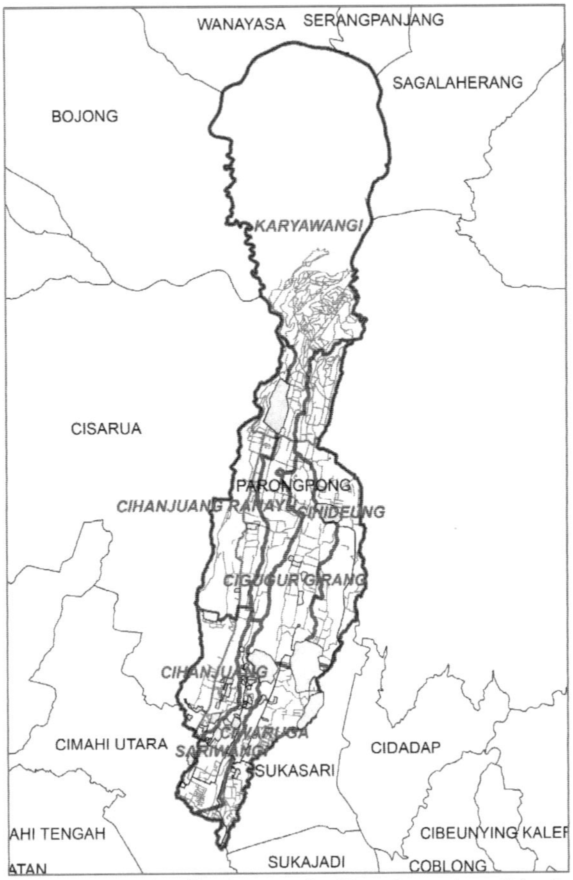

Figure 1. Distribution of Housing in District Parongpong. This housing which then become the object of study

2.2. Data Analysis

The analytical approach used is a quantitative analysis in which all the criteria and analysis undertaken can be measured, calculated, or operated mathematically. In more detail the method of analysis used are descriptive and association analysis.

Descriptive analysis is used to describe the existing condition of infrastructure in each housing. It will show what and how the infrastructure characteristics provided by developer. Association analysis is used to find out the related factors towards the infrastructure provisioning. Chi square analysis was used to identify the correlation between the condition of the existing infrastructure on housing characteristics and the factors. The correlation will indicate the presence of certain trends in the provision of infrastructure that is owned by the developer. In the study, chi square have been selected for this technique in accordance with the need to calculate the relationship between variables with nominal data types, which will only show the relationship between these variables.

Infrastructure examined in the study are clean water, sewage, drainage, and solid waste system. The factors that are expected have a significant correlation are the number of units, residential area, years of development, and housing classification by the target of market.

3. Result and Discussion

Discussion related research will be outlined according to the targets.

3.1. Basic Characteristics of Infrastructure Provision in Housing

Clean Water

Clean water is one of the top priorities in the provision of infrastructure in a residential neighborhood. Clean water is regarded as a vital necessity that must be met for the people. Before building a house is built, the existence of clean water resource will be confirmed first by the developer.

Based on the observation, all the entire housing have access to clean water, sourced from individual wells, artesian wells, springs, or PDAM provided by the developer. Clean water is supplied as part of housing. Developer also has the propensity to build a clean water network system by relying on ground water through individual wells or artesian wells. This indicates the desire of developers to create housing that is independent and also minimize resident dependence to the central urban network in terms of water supply.

Drainage

Drainage become one infrastructure which is quite important in a residential neighborhood. Drainage serves as a medium flow of runoff rainwater that falls to the ground. A good drainage will reduce the potential for inundation, especially if the housing area is at a location that is relatively low and has a high potential for inundation.

Despite the existence of a network of drainage channels are important, but not all housing have it. Based on observations made there were 88% of housing that has a good network of drainage channels with open drainage design, closed drainage, or a combination of both. A network of drainage channels have a high enough priority may be due to the ease and low cost required in development for the provision carried out simultaneously with the construction of the road network.

Sewerage

Sewerage provides a considerable influence on public sanitation problems in a residential neighborhood. Sewerage has a hierarchy of individual services to the communal. In the individual services, the infrastructure that must be provided is a septic tank while the communal infrastructure services that must be provided is a network of waste water treatment plan (WWTP).

Based on the observations made, it is known that the entire housing has had individual septic tanks and 88% of the housing of which have been equipped with a network of communal WWTP. This indicates that it is basically a network of sewerage infrastructure is definitely a basic infrastructure is provided to each house but the construction was not prioritized because it can be continued or handed over to the resident.

Waste Facility

The waste facility has the lowest priority to be provded than other basic infrastructure. Waste facilities including garbage containers and waste collection services are not regarded as something that is so important to be reserved in advance.

Based on the observations made, 80% of housing has garbage containers that can be used by resident to collect garbage produced every day in each home. 95% of housing has had a collection service for carrying garbage each house into further processing which is managed by the neighborhood association[3]. This indicates that the low priority the provision of the waste facility because it can not take place independently. There is a close relationship to the urban network. Waste management is continued process that taken a high cost. In the smallest units, waste management is the responsibility of government that is done by neighborhood association. Therefore, the provision of waste facilities can

[3] Neighborhood association (RT/RW) is the smallest unit of local government in Indonesia

be ruled out in advance because the operating system can further be delivered or undertaken in cooperation with surrounding territories.

The result of the observation made is showed in Table 2.

Table 2. Charateristic of housing infrastructure provided by developer

No	Element	Result
1	Drainage	
	Provided on the right and left of the road	51% housing has closed drain
		23% housing has open drain
		14% housing has mixed drain
		12% housing has none
2	Waste Facilities	
	Every house has individual waste facility, can be trash can or permanent garbage bin	80% housing has individual trash bin
		20% housing has none
	There is waste collection system	96% housing has waste collection facility
		4% housing has none
3	Clean Water	
	There is clean water in each house	65% housing has individual well
		20% housing get clean water from artesian well
		8% housing get clean water from water springs
		4% housing get clean water from PDAM
		2% housing get clean water from local PDAM
4	Sewerage	
	Every house has an access to individual or communal septic tank	100% housing has individual septic tank

3.2. Factors Related the Infrastructure Provisioning

Provision of basic infrastructure in a residential area is not significantly related by the number of housing units built and the construction of housing. Provision of basic infrastructure is significantly related by the type of housing that reflects the market target of each housing. Each type of infrastructure has a tendency to be provided on the type of housing. There is certain trends in the provision of infrastructure where the higher level of economic goals occupants will be more complete infrastructure components are built. The Table 3, Table 4, Table 5, and Table 6 show the correlation between infrastructure provision with years of construction, the total unit, and housing classification. The correlation of the factors and infrastructure provided is significantly can be seen in individual trash bin provision. With the significancy 0.048 is correlated with the year of construction and 0.001 is positive correlated with the classification of housing.

Table 3. Correlation between infrastructure provision and year of construction

No	Element (N=49)	Calculated X^2	df	Sig.	Table X^2	Result
1	Drainage	10.372	7	0.168	14.067	No correlation
2	Individual Trash Bin	14.211	7	0.048	14.067	There is correlation
3	Waste Collection	4.402	7	0.732	14.067	No correlation
4	Clean Water	-	-	-	-	Provided in all type of housing

No	Element (N=49)	Calculated X^2	df	Sig.	Table X^2	Result
5	Septic tank	-	-	-	-	Provided in all type of housing

Table 4. Correlation between infrastructure provision and total unit of housing

No	Element (N=49)	Calculated X^2	df	Sig.	Table X^2	Result
1	Drainage	7.898	9	0.544	16.919	No correlation
2	Individual Trash Bin	13.430	9	0.144	16.919	No correlation
3	Waste Collection	5.720	9	0.768	16.919	No correlation
4	Clean Water	-	-	-	-	Provided in all type of housing
5	Septic tank	-	-	-	-	Provided in all type of housing

Table 5. Correlation between infrastructure provision and total housing area

No	Element (N=49)	Calculated X^2	df	Sig.	Table X^2	Result
1	Drainage	44.347	47	0.583	67.505	No correlation
2	Individual Trash Bin	49.000	47	0.393	67.505	No correlation
3	Waste Collection	49.000	47	0.393	67.505	No correlation
4	Clean Water	-	-	-	-	Provided in all type of housing
5	Septic tank	-	-	-	-	Provided in all type of housing

Table 6. Correlation between infrastructure provision and classification of housing

No	Element (N=49)	Calculated X^2	df	Sig.	Table X^2	Result
1	Drainage	1.546	4	0.819	9.488	No correlation
2	Individual Trash Bin	18.340	4	0.001	9.488	There is a positive significant correlation
3	Waste Collection	7.139	4	0.129	9.488	No correlation
4	Clean Water	-	-	-	-	Provided in all type of housing
5	Septic tank	-	-	-	-	Provided in all type of housing

4. Conclusion

Storm water infrastructure, waste facilities and waste management services, clean water, and septic tanks have been generally available on housing built by developers. Variations encountered is the type of infrastructure provided. Only years of development and target the housing market are correlated with the provision of infrastructure, in this case the waste container. New construction housing and middle-high income housing above generally provide a trash receptacle individually. Condition of infrastructure provisioning in housing built by developers has not shown uniformity and so far there are no regulation related to the provision of infrastructure in built housing developer.

IOP Conf. Series: Earth and Environmental Science **124** (2018) 012012 doi:10.1088/1755-1315/124/1/012012

Acknowledgments
The authors would like to thank Institute Technology Bandung for financial support of this research under the scheme of Riset ITB 2016. The title of the research is 'Model Konseptual Penyediaan Infrastruktur berdasarkan Persepsi dan Preferensi Stakeholder'. Besides ITB, the authors would like to thank all the stakeholders involved in this program. However, the authors alone are responsible for any mistakes and shortcomings.

References

[1] Pemerintah Kabupaten Bandung Barat, "*Laju Pertumbuhan Penduduk Jabar Masih di Atas Standar Nasional Bandung Raya,*" 23 September 2015. [Online]. Available: http://www.bandungbaratkab.go.id/content/aju-pertumbuhan-penduduk-jabarmasih-di-atas-standar-nasional-bandung-raya. [Accessed 29 April 2016].

[2] Real Estate Indonesia for West Jawa Province, "*Data Perumahan Kecamatan Parongpong,*" West Bandung Regency, 2016.

[3] R. D. Woodson, *Be a Successfull Residential Land Developer*, New York: McGraw Hill, 2005.

[4] H. Winarso, "*Residential Land Developers Behaviour in Jabotabek, Indonesia,*" University College London, Inggris, 2000.

[5] T. Firman, "*New town development in Jakarta Metropolitan Region: a perspective of spatial segregation,*" Habitat International 28, pp. 349-368, 2004.

[6] "*Data Prasarana, Sarana, dan Utilitas (PSU) Kecamatan Parongpong,*" Kabupaten Bandung Barat, 2009.

IOP Conf. Series: Earth and Environmental Science **124** (2018) 012013 doi:10.1088/1755-1315/124/1/012013

Development Study of Pedestrian Bridge at Gramedia Bookstore Jalan Raden Intan Bandar Lampung

C M Bernaditha

Civil Engineering Department, Institut Teknologi Sumatera
Jl. Terusan Ryacudu, Lampung Selatan, Lampung, Indonesia

ABSTRACT. Bandar Lampung with high enough population densities has provides transportation facilities for pedestrian such as pedestrian bridge. This pedestrian bridges spread at Bandar Lampung's traffic congested area, shopping centre nor education centre. Jl. Raden Intan as one of primary collector road with four lanes one direction at Bandar Lampung has high LHR (average daily traffic) movemenet pattern especially at morning, day and afternoon rush hour that make it difficult for pedestrian who want to cross the road. Therefore pedestrian bridge at this section Jl. Raden Intan highly needed especially at in front of Gramedia Bookstore with large amount of crossing pedestrian volume. From this research and analysis, found that number of LHR (average daily traffic) at Jl. Raden Intan shows large number traffic volume that is 4509 passenger car unit/hour at morning rush hour (07.00-08.30), with value of V/C Ratio or Degree of Saturation reach 0,92 (E category), while the amount of pedestrian who cross ahead from Gramedia Bookstore to Bank Muammalat is 29 people per 15 minutes. Other than that based on the calculation results of pedestrian volume and traffic volume at rush hour as follow: average pedestrian volume at rush hour is 146 people/hour between the range 100-1250 people/hour and traffic volume 7521 vehicles/hour over than 7000 vehicles/hour, and also the value $PV^2=1{,}682x10^{10}$ which is means the value of PV^2 worth over $2x10^8$, moreover the speed plan Jl. Raden Intan between 60-80 km/hour above 70 km/hour. Based on the calculation and analysis above, it can be concluded transportation facilities recommended for Jl. Raden Intan is pedestrian bridge.

1. Introduction

The population of Bandar Lampung reached 960.695 inhabitans with density of population 5678 people per km2 (Central Statistics Agency / *BPS* of Lampung Province, 2014). As the developing city, Bandar Lampung has transportation problems one of which is provision of supporting facilities for pedestrian who meet road safety.

Jl. Raden Intan as one of primary collector road with four lanes one direction and speed plan 60-80 km/hour, so has high Average Daily Traffic movement system especiallyat morning, day and afternoon rush hour that make it difficult for pedestrian who want to cross the road. For those reasons, the pedestrian bridge at Jl.Raden Intan is highly needed especially with large amount of crossing pedestrian volume.

Pedestrian bridge function as mode transfer walking pedestrian to crossing pedestrian, around area with very high vehicle movement. Therefore, pedestrian bridge as safety alternative avoiding traffic accident as well traffic congestion since zebra cross and pelican are not able to solve the pedestrian problem well.

2. Literature Review

This section provides an overview of available Indonesia globally study on some important perspective of pedestrian operations in urban networks, and also based on the type of crossing means.

Pedestrian behaviour in urban networks was described by Hoodendoorn and Bovy (2004) in Zheng and Elefteriadou (2016) as hierarchical structure with: strategic level (departure time choice); tactical level (activity scheduling and route choice); and operational level (road crossing and interactions). The tactical decision interacts with the operational level when, for example, pedestrian travel route may change due to available crossing facilities, and pedestrian crossing location may affect the pedestrian overall travel time. This structure explains the relationship among these three levels and emphasizes the necessity for an integrated method for pedestrian operation analysis. However, most existing studies ignore these mutual impacts, and pedestrian travel time is typically analyzed only at the intersection level. Thus, in order to approximate the pedestrian perspective, it is necessary to develop a pedestrian travel time model that considers the entire trip.

2.1 Crossing Pedestrian Facility

Crossing pedestrian facility is one of pedestrian facilities on the highway to concentrate pedestrian who want to cross the road. Ideally all road crossings using type of crossing separate from road, which means there is no possibility conflict between pedestrian and vehicle (Susilo, 1984 in Andreas, 2012).

Based on Decree of the Land Transportation General No: SK.43/AJ 007/DRJD/97 about Pedestrian Facilities Engineering in the City Area, crossing pedestrian facility divided into 2 (two) levels which are one level crossing and interchange crossing.

a. One Level Crossing
One level crossing for pedestrian consists of zebra cross with protector, zebra cross without protector, pelican with protector and pelican without protector
b. Interchange Crossing
Interchange crossing for pedestrian consists of pedestrian bridge and pedestrian tunnel.

2.2 One Level Crossing

Selection criteria for one level crossing:

a. Based on the empirical formula (PV^2), where P is the flow of pedestrian crossing along 100 meter per hour (people/hour), and V is the flow of vehicle per hour in 2 (two) directions (vehicle/hour)
b. The value of P and V is the average flow of pedestrian and vehicle in 4 (four) peak hours a day with initial recommendations such as Table 1.

Table 1. Recommendation of crossing pedestrian selection

PV^2	Pedestrian Volume (P) (people/hour)	Vehicle Volume (V) (vehicle/hour)	Type of Facilities
$>10^8$	50 - 1100	300 - 500	*Zebra Cross (ZC)*
$>2 \times 10^8$	50 - 1100	400 - 750	ZC with protector
$>10^8$	50 - 1100	>500	*Pelican* (P)
$>10^8$	>1100	>300	*Pelican* (P)
$>2 \times 10^8$	50 - 1100	>750	*Pelican* with protector
$>2 \times 10^8$	>1100	>400	*Pelican* with protector

Source: Directorate General of Bina Marga, *Pedestrian Facilities Engineering in the City Area* (1997)

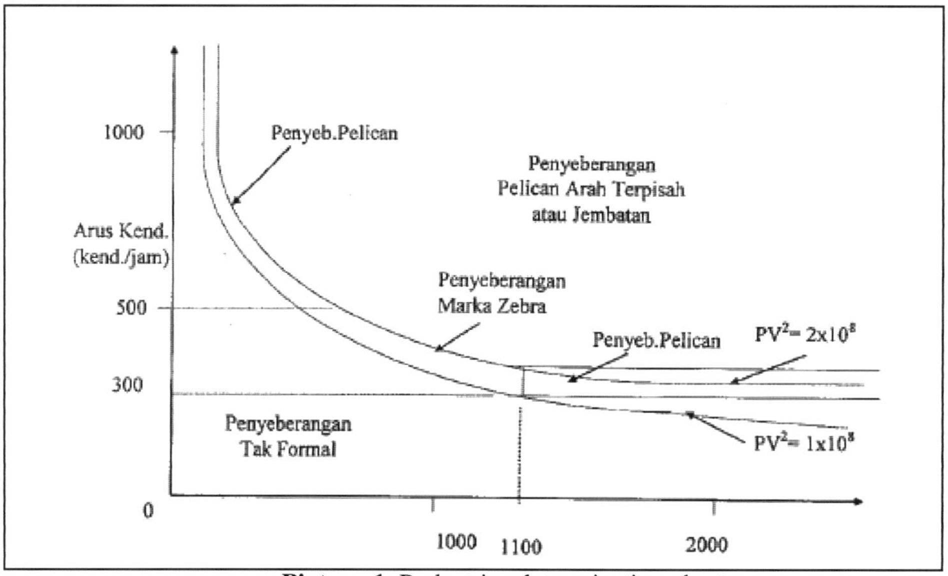

Picture 1. Pedestrian determination charts

Source: Directorate General of Bina Marga, *Pedestrian Facilities Engineering in the City Area* (1997)

2.3 Interchange Crossing

Selection criteria for interchange crossing:

a. Based on the empirical formula (PV^2), where P is the flow of pedestrian crossing along 100 meter per hour (people/hour), and V is the flow of vehicle per hour in 2 (two) directions (vehicle/hour)
b. The value of PV^2 more than $2x10^8$, pedestrian flow (P) more than 1100 people/hour, the flow of vehicle in 2 (two) directions value (V) more than 750 vehicle/hour, which taken from 4 (four) peak hours.
c. On the road with the speed plan 70 km/hour
d. On the strategic area with no possibility for pedestrian to crossing but through pedestrian bridge

Factors to be considered in the use of interchange crossing according to Peter Bottomley (in Amalia L, 2005)

a. Level of safety and secure factor to avoid accident occurrence
b. Calculation quantitatively of conflict between pedestrian and vehicle
c. Cost efficiency
d. Use accuracy of crossing pedestrian facility especially interchange crossing in term of design and location, as well as comfort and convenience using

Selection criteria for interchange crossing based on Tanan and Sailendra (2011) from their book '*Training Module for Technical Planning of Pedestrian Facilities*' is the value of P flow of pedestrian crossing in each hour reach over >1100 people/hour, with the value of V flow of vehicle per hour reach over >750 vehicle/hour and the value of PV^2 reach over $>2x10^8$. From these variety of values can be recommended interchange crossing.

Furthermore explained the requirements of pedestrian bridge are:

a. Technical condition of crossing bridge construction follow decree of planning for pedestrian bridges for pedestrian in urban area
b. Pedestrian crossing bridge is a building for crossing pedestrian from side to other side cross section of the road. Pedestrian bridge must be build with strong construction and easy to maintain
c. Pedestrian bridge has minimum width 2 meters with maximal ladder gradient 20°
d. If pedestrian bridge also for bicycles then minimum width is 2,75 meters

e. Pedestrian bridge must be equipped with an adequate fence
f. At the centre of the bridge must be equipped with slab which can be used as a wheelchair facility for disabilities
g. Location and bridge building must be in accordance with the needs of pedestrians and as well as aesthetic
h. Pedestrian bridge placement should not reduce the effective width of sidewalk

While the requirements for pedestrian tunnel are as follows:
a. The technical provision of tunnel pedestrian follow the specification of tunnel technique
b. The tunnel must be equipped with adequate lightning. The specification for placement of lighting will be arranged in a separate document
c. Pedestrian tunnel has minimum width 2,5 meters. If the tunnel also for bicycle then minimum width is 2,75 meters
d. Minimum height of pedestrian tunnel is 3 meters.

3. Research Methods

Regional study directed in accordance with the problems identification, be concentrated at Jl. Raden Intan especially in front of Gramedia Bookstore. This is because at rush hour traffic flow increases making it difficult for pedestrians to cross the street.

3.1 Primary Data

Primary data collection is done directly by traffic counting survey at study location, with the purpose of obtaining accurate information related with road network traffic performance around study location. Beside the traffic counting, another survey are also done such as crossing pedestrian survey and geometry traffic survey.

The point of survey area assigned at JL.Raden Intan in front of Gramedia Bookstore, namely survey point "S" by placing four surveyor at the loaction.

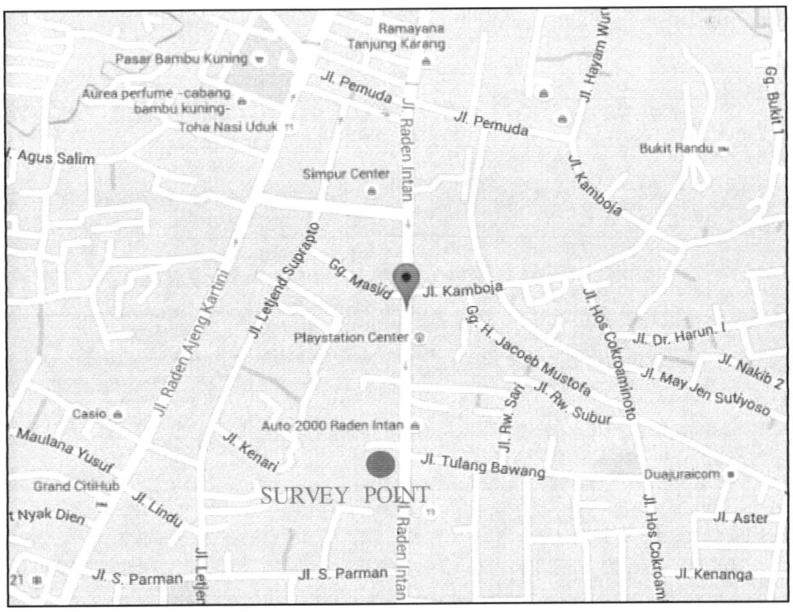

Picture 2. Survey location for traffic counting and pedestrian crossing

3.2 Secondary Data

For this current research, the secondary data given of Spatial Plan of Bandar Lampung (*Rencana Tata Ruang Wilayah RTRW)* Year 2011-2030, issued by Bappeda Kota Bandar Lampung. Furthermore secondary data in Bandar Lampung in Figures 2014, issued by BPS and Bappeda. In addition, secondary data required next is the road network map Bandar Lampung in Auto Cad format, which was issued by Bappeda Lampung Province.

The implementation procedures for survey and data processing stage starting with the implementation survey procedure: the form and equipment preparation, conducting survey, collecting survey result. Followed by a stage of data compilation: volume of crossing pedestrian, volume of vehicles, speed vehicle.

The next stage is data analysis stage which aims to analyse the problems that exist for the proposed solutions and consideration of the impact of future problems: counting PV^2, crossing pedestrian (P) and vehicle volume (V), road capacity, speed vehicle, the characteristics of pedestrians.

After done data analysis stage, the next is test phase of data analysis which purpose whether there is a relationship between a pedestrian crossing with traffic passing on the road crossings are reviewed as proposed solutions to problems.

The last stage is the step of determining conclusions: giving the use of conformity assessment crossing facilities, and giving the use of conformity assessment of pedestrian facilities to be built.

3.3 Survey Data

The preliminary survey to collect geometry data been done the day before the traffic counting survey, the information given include road wide effective, length of the segment, lane width, side walk.

The data traffic volume held on Thursday, July 21 2016 for 6 hours representative morning, day and afternoon busy hours (06.30-08.30, 11.00-13.00, 16.00-18.00).

The condition of the average daily traffic at Jl.Raden Intan shows the volume of traffic is more dense in the morning (07.30-07.40) compare to during another peak hour that is 792 pcu/10 minutes or 4509 pcu/hour. The density of activity Jl. Raden Intan can be understood since Jl.Raden Intan is the axis that connects to the central government and trade center. Average daily traffic hour Jl.Raden Intan 2943 pcu/hour. This condition can be understood since Jl. Raden Intan is the backbone of urban roads in Bandar Lampung, liaison area of residence, place of activity and education area.

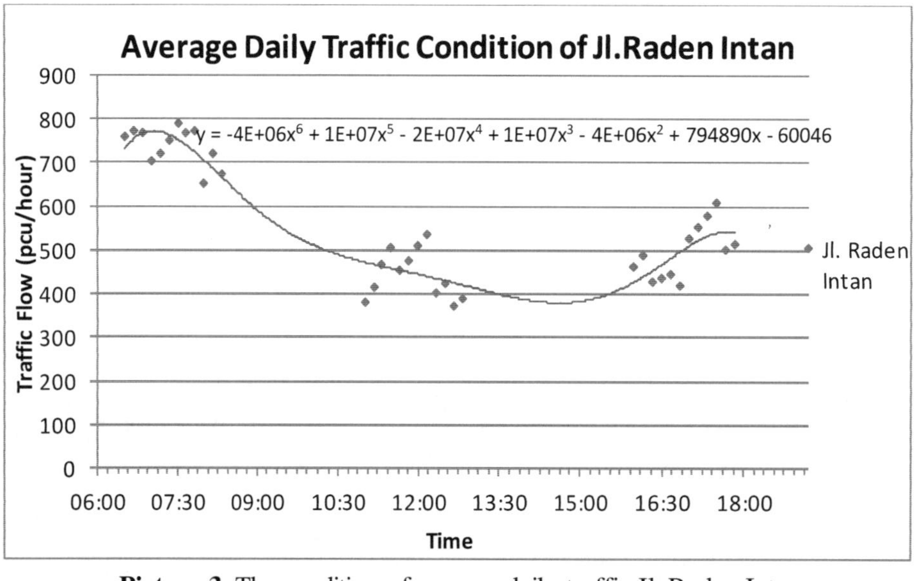

Picture 3. The condition of average daily traffic Jl. Raden Intan
Source: Analysis results, 2016

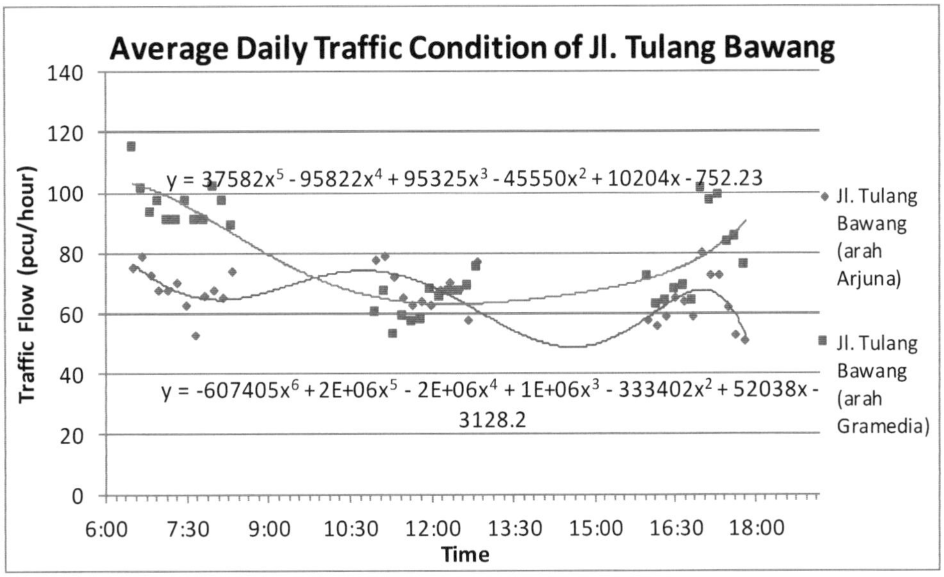

Picture 4. The condition of average daily traffic Jl. Tulang Bawang
Source: Analysis results, 2016

Average daily traffic condition Jl. Tulang Bawang for total 2 directions towards Arjuna neither Gramedia intersection, show higher traffic volume in the peak hour at morning season (06.30 – 06.40) compare to peak hour at noon season, which is 190 pcu/10 minutes, maximum traffic volume towards Arjuna reach the highest level 115 pcu/10 minutes at approximately time between 06.30 – 06.40.

The cause of the traffic flow density Jl. Tulang Bawang is an alternative access from Jl. Raden Intan to Jl. HOS Cokroaminoto either Jl. Jendral Sudirman, as the center of trip generation.

3.4 Pedestrian Data

Crossing pedestrians data is calculated at 15 minute intervals during the two hours represent the time morning, day and afternoon busy hour. Crossing pedestrian data be divided into 2 (two) groups there are crossing pedestrian from Muammalat Bank into Gramedia Bookstrore and crossing pedestrian from Gramedia Bookstore into Muammalat Bank.

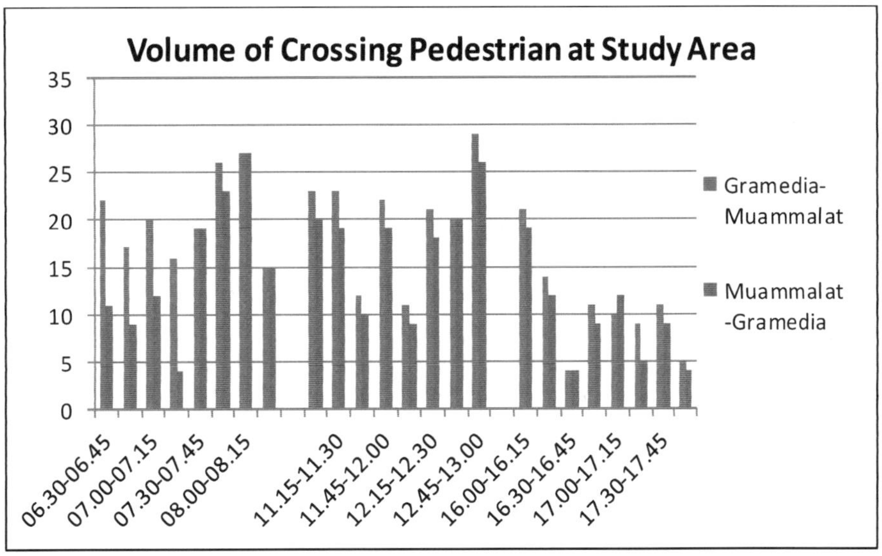

Picture 5. The volume of crossing pedestrian at Jl. Raden Intan on Thursday, July 21 2016
Source: Analysis results, 2016

3.5 Urban Roads Performance Based on HCM

The Highway Capacity Manual or in Indonesia we used to call it Manual Kapasitas Jalan Indonesia (MKJI, 1997) provides guidelines, concepts and computational procedure in highway facilities, including highway, urban road, interurban road, signalized and unsignalized intersection, pedestrian and bicycle on the performance on these systems. The most current edition of HCM provides several methodologies for evaluating the pedestrian Level of Service (LoS) of the different urban street facilities.

Urban roads's Level of Service (LoS) or urban roads performance in the area study calculated and analysed based on the value of capacity, degree of saturation, travel speed, free flow and also travel time. The results of the performance calculation roads for rush hour each morning, afternoon and evening given below.

Table 2. Calculation of urban roads performance Jl. Raden Intan based on HCM

Time	Volume (vehicle/hour)	Flow (pcu/hour)	Capacity (pcu/hour)	Degree of Saturation	Velocity (km/jam)
06.30-07.30	9795	4389	4765	0,92	34,3
07.30-08.30	9917	4316	4765	0,91	34,5
11.00-12.00	5224	2629	4765	0,55	45,0
12.00-13.00	5148	2582	4765	0,54	45,1
16.00-17.00	5390	2629	4765	0,55	45,0
17.00-18.00	6970	3232	4765	0,68	42,5

Source: Analysis results, 2016

The value of Jl. Raden Intan's V/C Ratio reach the lowest level at 0.92 on morning peak hour 06.30-07.30, with the highest vehicle velocity 45,1 km/jam. This road performance shows that Jl.Raden Intan as the main road access to connect residence area to place of activity, central government and education area. This also caused pedestrians at Jl. Raden Intan find difficulty to cross the road since the dense vehicle so high that no opportunities for pedestrians to cross.

3.6 Recommendations of Pedestrian Bridge

After learning in detail about the volume and patterns movement and performance of urban roads in the study area using MKJI, also crossing pedestrian data from Muammalat Bank to Gramedia Bookstore and reverse direction, next could be synergized data that exists as a decisive criterion in choosing a crossing facility that is used by engineering facilities in the Region Walking City DPU issued by Directorate General of Highways.

1. **P** is the flow of crossing pedestrians along 100 m each hour (pedestrian /hour) and V is vehicle flow each hour
2. **P** and **V** are an average flow of pedestrians and vehicle at 4 rush hour. So that P and V are summed every hour to obtain empirical formula PV^2.

Table 3. A review on the bridge crossing facilities Jl. Raden Intan on Thursday, July 21 2016

Time	P (people/hour)	V two directions (kend/jam)	P.V2	4(P.V2) maximum
06.30-07.30	111	9795	1,065E+10	1,065E+10
07.30-08.30	171	9917	1,682E+10	1,682E+10
11.00-12.00	148	5224	4,039E+09	4,039E+09
12.00-13.00	154	5148	4,081E+09	4,081E+09

Time	P (people/hour)	V two directions (kend/jam)	P.V2	4(P.V2) maximum
16.00-17.00	94	5390	2,731E+09	
17.00-18.00	65	6970	3,158E+09	

Source: Analysis results, 2016

To recommend crossing facilities in accordance with existing requirements can be calculated from the volume of pedestrian and the highest volume of vehicle of peak hours. Which is when the average value of 4 PV^2 largest such calculations below:

$$P = (111 + 171 + 148 + 154) / 4$$
$$= 146 \text{ pedestrian/hour}$$
$$V = (9795 + 9917 + 5224 + 5148) / 4$$
$$= 7521 \text{ vehicle/hour}$$

Then the average number of pedestrians at rush hour is 146 pedestrians/hour, between 100-1250 pedestrians/hour. The amount of vehicle volume 7521 > 7000 pcu/hour, and produced $PV^2 = 1.682E+10$ which is mean $PV^2 > 10^{10}$.

Based on the walking engineering facilities in the urban roads territory issued by DPU The Directorate General of Highway, can be deduced that the facilities recommended using the **Pedestrian Bridge.**

4. Conclusion and Recommendation

From this research and analysis, found that number of LHR (average daily traffic) at Jl. Raden Intan shows large number traffic volume that is 4509 passenger car unit/hour at morning rush hour (07.00-08.30), with value of V/C Ratio or Degree of Saturation reach 0,92 (E category), while the amount of pedestrian who cross ahead from Gramedia Bookstore to Bank Muammalat is 29 people per 15 minutes. Other than that based on the calculation results of pedestrian volume and traffic volume at rush hour as follow: average pedestrian volume at rush hour is 146 people/hour between the range 100-1250 people/hour and traffic volume 7521 vehicles/hour over than 7000 vehicles/hour, and also the value $PV^2=1,682x10^{10}$ which is means the value of PV^2 worth over $2x10^8$, moreover the speed plan Jl. Raden Intan between 60-80 km/hour above 70 km/hour. Based on the calculation and analysis above, it can be concluded transportation facilities recommended for Jl. Raden Intan is pedestrian bridge.

This research only do the traffic counting and pedestrians crossing at 6 (six) hours consisting of 2 (two) peak hours at morning, 2 (two) peak hours at noon and another 2 (two) peak hours at evening. For better results on next research, the survey shall do 12 (twelve) hours nonstop from early morning (06.00) to late evening (18.00) in order to obtain the real movement pattern.

For effective pedestrian bridge use, shall do the following like widening the road pavement and build up a fence on the median. Lack of side barriers and traffic smooth causing increased vehicle speed so it will complicate pedestrians to cross the road, hopefully will make crossing pedestrian using pedestrian bridge.

5. References

[1] Departemen Pekerjaan Umum (Juni 1997). "Manual Kapasitas Jalan Indonesia". Departemen Pekerjaan Umum Direktorat Jenderal Bina Marga.

[2] Amalia, Listiati (2005). "Kajian Efektifitas Jembatan Penyebrangan Pejalan Kaki pada Pusat Perdagangan di Kota Semarang". Tesis. Program Magister Teknik Sipil. Universitas Diponegoro. Semarang

[3] Andreas, Richard (2012). "Studi Efektifitas Jembatan Penyebrangan (Studi Kasus: Jalan Sisingamangaraja Medan)". Tugas Akhir. Program Pendidikan Ekstension. Departemen Teknik Sipil. Fakultas Teknik. Universitas Sumatera Utara. Medan

[4] Departemen Perhubungan (1997). "Keputusan Direktur Jenderal Perhubungan Darat Nomor: SK.43/AJ 007/DRJD/97 Perekayasaan Fasilitas Pejalan Kaki di Wilayah Kota". Departemen Perhubungan. Direktorat Jenderal Perhubungan Darat. Jakarta

[5] Direktorat Jenderal Bina Marga (2010). "Rencana Strategis 2010 – 2014 Direktorat Jenderal Bina Marga". Direktorat Jenderal Bina Marga. Kementrian Pekerjaan Umum

[6] Kota Bandar Lampung (2011). "*Peraturan Daerah Kota Bandar Lampung Nomor 10 Tahun 2011 tentang Rencana Tata Ruang Wilayah Tahun 2011-2030*". Pemerintah Daerah Kota Bandar Lampung. Lampung.

[7] Pemerintah Kota Bandar Lampung (2014). "*Bandar Lampung Dalam Angka 2014*". Badan Pusat Statistik dan Badan Perencanaan Pembangunan Daerah. Bandar Lampung.

[8] Tanan, Natalia dan Sailendra, A.B. (2011). "Modul Pelatihan Perencanaan Teknis Fasilitas Pejalan Kaki". Kementrian Pekerjaan Umum. Puslitbang Jalan dan Jembatan. Bandung.

[9] Zheng Y, Elefteriadou L, Chase T, Schrieder B and Sisiopiku V (2016). "*Pedestrian Traffic Operation in Urban Networks*". ISEHP 2016. International Symposium on Enhancing Highway Performance. Volume 15, 2016. Pages 137-149

Study on Performance of Intersection Around The Underpass Using Micro Simulation Program

J Arliansyah[1], R T Bawono[1]

[1] Department Of Civil Engineering, University of Sriwijaya, Jl. Raya Palembang-Prabumulih KM 36 Inderalaya, Sumatera Selatan, Indonesia

Abstract. In order to overcome the problems of congestion at the major intersections in Palembang City, there have been grade separation constructions in the form of flyover or underpass. One of them is the intersection of Patal Pusri Underpass. The smooth traffic due to the underpass construction needs to be balanced by the arrangement management at the intersections located close to the underpass to get the best network performance since both affect each other performance. The Taman Kenten and Seduduk Putih junctions which are just 569 and 226 meters from the underpass of Patal Pusri has to be analyzed for its needs of management and traffic arrangements to reduce its impact on the performance of the intersection of Patal Pusri or vice versa. Some alternatives of management and traffic arrangements were developed to get the best solutions and the Vissim 8.00 microsimulation program was used to evaluate the performance of intersections in the network where the compared parameters are the value of the queue length and average delay. The results of the analysis and the conducted modeling show that the best solution to optimize the performance of the two junctions are to make geometric changes and diversion of traffic flow at Seduduk Putih Junction and making geometric changes at Taman Kenten Junction.

1. Introduction

Palembang City as a metropolitan city and a capital city of South Sumatra Province is a center of administration, commerce, industry and settlement in South Sumatra. The number of population of Palembang City in 2016 has reached about 1.8 million people that need quite a lot of the existing movement, and traffic congestion is a very serious problem in the city.

One of the solutions of overcoming the traffic congestion at the intersections of the city is by constructing flyovers or underpasses, one of which is the underpass of Patal Pusri

To optimalize the arrangement of traffic at the underpass of Patal Pusri, it requires to perform a traffic management nearby the Underpass of Patal Pusri, that is the Junctions of Taman Kenten and Seduduk Putih whose distances are within 569 and 226 meters from the junction of Patal Pusri.

In analyzing the traffic performance, the Software Vissim was used and several alternative traffic arrangements at the existing junctions were developed in this study.

2. Literature Review

Microsimulation traffic model is a device to simulate the individual vehicle behavior in the road network to estimate possible impacts of traffic pattern change resulted from the traffic flow change or from changes in the physical environment.

Microsimulation model also has a capability of simulating queuing conditions and providing results of density rate and road service. This capability makes this type of model very useful for analyzing traffic operations in urban areas and the city center including interchanges, roundabouts, not signalized and signalized intersections, signal coordinated corridors, and area networks. Microsimulation even reflects relatively small changes in the physical environment such as narrowing lanes, relocating the stop line before the intersection, or optimizing operation of toll booths.

IOP Conf. Series: Earth and Environmental Science **124** (2018) 012014 doi:10.1088/1755-1315/124/1/012014

There are two methods of simulation process in Vissim, namely Static Vehicle route and Dynamic Assignment, in which the Dynamic Assignment is more used for simulation at a network because route choice highly depends on the traffic conditions that take place. The writer use VISSIM Manual as a guide to made the model simulation [1].
Several studies successfully applying the Vissim program have been reported. [2, 3, 4, 5, 6]

3. Methodology

The locations of the study were the underpass of Patal junction and two nearby junctions, that is the Junctions of Taman Kenten and Seduduk Putih. The locations are shown in Figure 1.
The methodological procedure is described as follows:

1. Data collection
2. Modelling contained the above three junctions using Vissim program,
3. Calibration and validation models based on GEH Statistic that revealed by F Olga (2012) Formula 1. Shows the GEH Statistic that developed by Olga [7]

$$\text{GEH} = \left(\frac{2x(qsimulated - qobserved^2)}{(qsimulated + qobserved)}\right)^{1/2} \qquad (1)$$

Table 1. GEH Statistic Criteria

Condition	Remark
GEH ≤ 5,0	Accepted
5,0 ≤ GEH ≤ 10,0	Posibility some error in model
GEH ≥ 10,00	Rejected

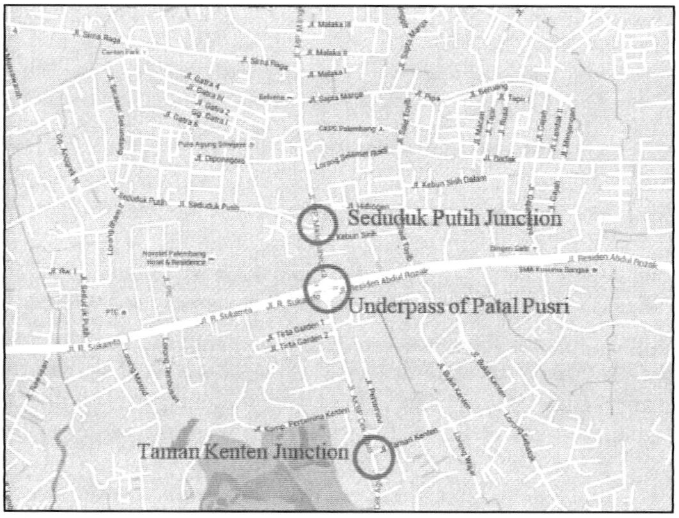

Figure 1. Locations of Study.

4. Evaluation of the existing conditions at the Junctions of Taman Kenten and Seduduk Putih.
5. Development of alternative actions for improvement are as follows:
   - Geometric change, performed both at Taman Kenten Junction and Seduduk Putih Junctions. Figures 2 and 3 show additional widths road to avoid some queue at intersection in congestion condition.
   - The traffic diversion is possibly conducted at Seduduk Putih Junction where the minor traffic flow from Seduduk Street is diverted to other routes.

- The installation of traffic signals can be carried out at the two junctions. The cycle time for each junction is shown at Figure 4 and 5
- Combination of the above alternatives.

6. Evaluation of the junction performance of all alternative improvements is conducted by using Vissim program.

The study location at Underpass Patal Pusri Intersection and other intersections around it include the Seduduk Putih and Taman Kenten junctions located 226 and 569 meters from underpass Patal Pusri. The coordinate location based on UTM at 2°57'00.7"S 104°46'04.4"E. This location known as major intersection in Palembang City, connected R Soekamto Street at west, Residen Abdul Rozak street at east, MP Mangkunegara Street at north, and AKBP Cek Agus at South which is all of the street is line bunding to support of palembang citizen movement. This underpass was constructed in 2014 to increase the level of service at Patal Pusri intersection.

The Seduduk Putih and Taman Kenten junctions got the influence due to the smooth traffic flow at Underpass Patal Pusri. The traffic flow at these junctions needs to manage to get best performance in order to increase the level of service at these junctions and underpass Patal Pusri.

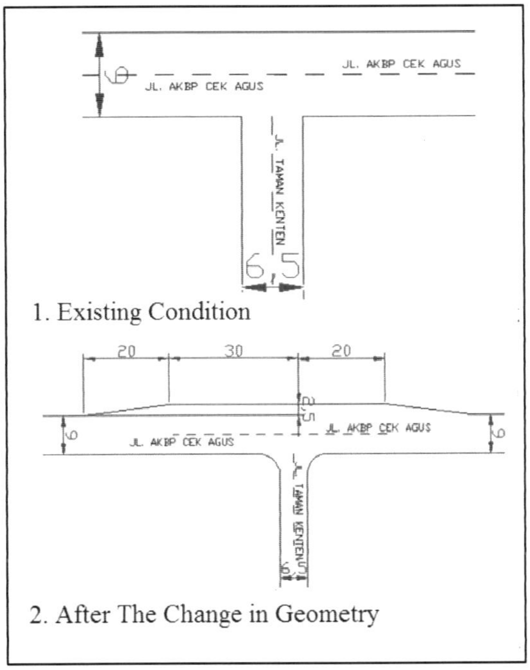

Figure.2. Geometric Change of Taman Kenten Junction

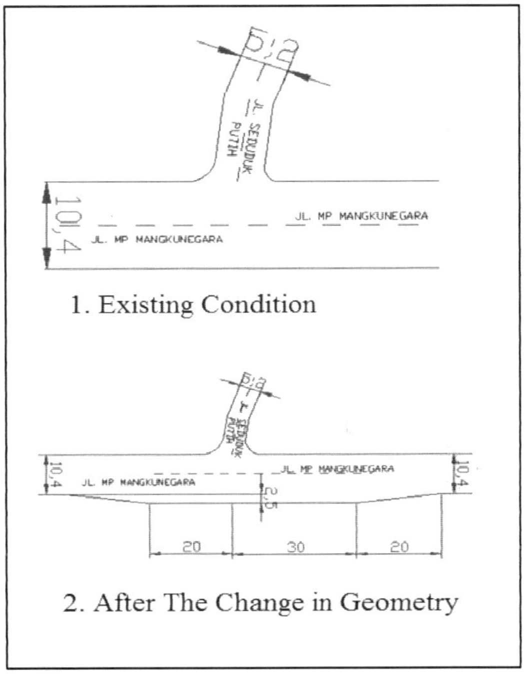

Figure.3 Geometric Change of Seduduk Putih Junction

Both of Figure 2 and Figure 3 ilustrate one alternative development at both junctions providing an additional width of 2.5 m and length of 70 m near the intersection that can be used when some congestion occurs at the intersection so the vehicles can use the additional lane to continue their movement.

IOP Conf. Series: Earth and Environmental Science **124** (2018) 012014 doi:10.1088/1755-1315/124/1/012014

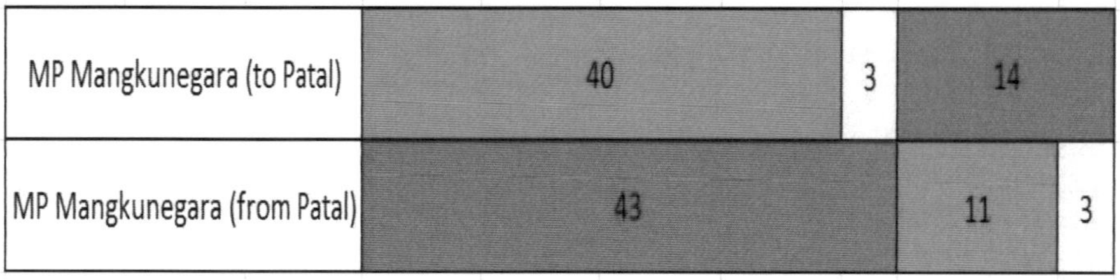

Figure.4 Cycle time at Seduduk Putih Junction

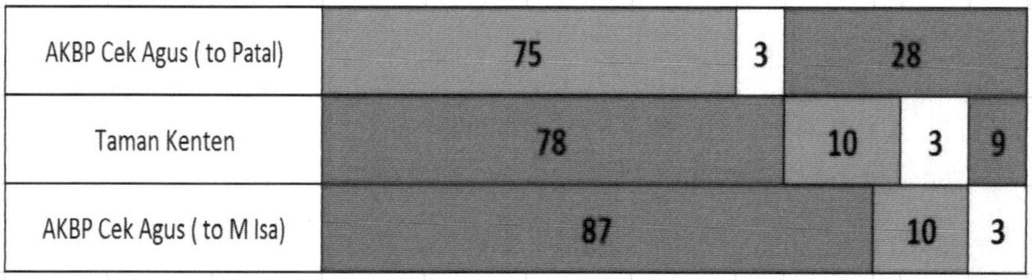

Figure 5. Cycle Time at Taman Kenten Junction

4. Data Presentation Analysis and Discussion

This section describes about the data collection, modelling the traffic condition in VISSIM 8.00 Program, calibration of the model based on GEH Statistic, and analysis of the traffic performance of the solution alternatives at both junctions around underpass Patal Pusri. The best alternative was selected among the alternatives developed.

4.1. Data Collection

The data need to be collected for analysis in this study are traffic volume, free flow speed and existing road geometric.

4.2. Calibration of theModel

Traffic flow data collected at analyzed intersections were used to calibrate microsimulation models parameters in order to reach reliable and realistic outputs. In the calibration process, traffic volume in peak hour was compared with the results of traffic volume from VISSIM program. GEH statistic that mention in formula 1 was used.

Table 2. Calibration Analysis of Seduduk Putih Junction

No	Section	Number of Vehicle Observed	Number of Vehicle based on Simulation	GEH Value	Remark
1	MP Mangku Negara (to Patal)	1810	1785	0.589	accepted
2	Seduduk Putih	330	345	0.816	accepted
3	MP Mangku Negara (to Kenten)	2075	2002	1.617	accepted
	Average GEH Value			1.008	accepted

Table 3. Calibration Analysis of Seduduk Putih Junction

No	Section	Number of Vehicle Observed	Number of Vehicle based on Simulation	GEH Value	Remark
1	AKBP Cek Agus (To Patal)	2037	1993	0.980	accepted
2	Taman Kenten	367	384	0.877	accepted
3	AKBP Cek Agus (To M Isa)	1947	1902	1.026	accepted
	Average GEH Value			0.961	accepted

Based on Table 2 and Table 3, it is found that the traffic flow resulted in VISSIM model simulation can be accepted so the simulation model is ready to be used to analyze the scenario developed.

4.3. Seduduk Putih Junction

Seduduk Putih Junction located 226 m at north Underpass Patal Pusri intersection and the traffic volume at this junction shown in Table 2. At this junction, six solution alternatives were developed to solve traffic problems. Table. 4 shows the developed alternatives combined with the performance of Seduduk Putih Junction.

Table 4. Performance of Seduduk Putih Junction

No	Alternative	Street	Queue Lenght (m)	Delay (s/veh)
1	Existing	MP Mangkunegara (to patal)	180.58	184.68
		Seduduk Putih	112.74	
		MP Mangkunegara (to kenten)	211.18	
2	Geometric Change	MP Mangkunegara (to patal)	117.86	110.89
		Seduduk Putih	75.2	
		MP Mangkunegara (to kenten)	191.68	
3	Geometric Change Combine with Diversion Traffic Flow	MP Mangkunegara (to patal)	26.75	76.4
		Seduduk Putih	0	
		MP Mangkunegara (to kenten)	190.46	
4	Installation of traffic light	MP Mangkunegara (to patal)	175.94	80.27
		Seduduk Putih	73.27	
		MP Mangkunegara (to kenten)	221.23	
5	Geometric change combine with installation traffic light	MP Mangkunegara (to patal)	125	78.26
		Seduduk Putih	71.21	
		MP Mangkunegara (to kenten)	191.23	
6	Geometric change combine with installation traffic light and diversion traffic flow	MP Mangkunegara (to patal)	125.91	91.64
		Seduduk Putih	0	
		MP Mangkunegara (to kenten)	190.24	

Based on Table 4, it is found that the third alternative is the best scenario among the scenarios developed because it has the shortest queue length while the delay is 76.4 second. The zero value of flow at seduduk putih street happen because the vehicle flow is diversed to use other route.

4.4. Taman Kenten Junction

Taman Kenten Junction located 569 m at south Underpass Patal Pusri intersection. Total traffic volume at this junction is 896 vehicle at peak hour. At this junction four solution alternatives was developed to handle traffic problems. Table 5 shows the developed alternatives combined with the performance of Taman Kenten Junction.

Table 5. Performance of Taman Kenten Junction

No	Alternative	Street	Queue Lenght (m)	Delay (s/vehicle)
1	Existing	AKBP Cek Agus (to patal)	504.72	135.77
		Taman Kenten	131.05	
		AKBP Cek Agus (to M Isa)	256.08	
2	Geometric Change	AKBP Cek Agus (to patal)	504.72	76.15
		Taman Kenten	78.21	
		AKBP Cek Agus (to M Isa)	196.71	
3	Installation of traffic light	AKBP Cek Agus (to patal)	504.38	89.02
		Taman Kenten	74.81	
		AKBP Cek Agus (to M Isa)	211.71	
4	Geometric change combine with installation traffic light	AKBP Cek Agus (to patal)	530.22	86.41
		Taman Kenten	117.85	
		AKBP Cek Agus (to M Isa)	235.84	

Based on Table 5, we can see that the second alternative is the best scenario among the other alternatives developed, because it has the shortest queue length while the delay is 76.15 second.

Based on the above cases, all solution alternatives can improve the service performance of the junctions however the geometric changes and traffic flow diversion are the best alternative among the alternatives developed.

5. Conclusion

In order to improve the service performance of the junctions studied, several improvement alternatives which are geometric changes, traffic flow diversion, traffic light installments and their combinations was developed. The result indicated that the performance of Seduduk Putih Junction can be improved by using combination of geometric changes and traffic flow diversion. Geometric changes is recommended alternative to improve the performance of Taman Kenten Junction.

6. References

[1] *PTV Vissim 8 User Manual* 2015, (Germany: PTV AG)

[2] Della RH, Hanafiah, Arliansyah J and Artiansyah R 2015 Traffic performance analysis of u-turn scenario; a case study at Soekarno Hatta Road, Palembang, Indonesia *Procedia Eng.* **125** pp. 461–6

[3] Lin D, Yang X and Gao C 2013 VISSIM-based simulation analysis on road Network of CBD in Beijing China *Procedia Soc. Behav. Sci.* **96** pp. 461-472

[4] Yang Y and Yang G 2016 Study of Intersection Optimization Near Transportaion Hub Based on Vissim *Int. J. Sig. Process. Pattern Recognit.* **9** pp. 323–32

[5] Zargari S A, Dehghani N and Moraveji G 2014 Optimal signal control in urban road networks with high priority congested centers *Int. J. Transp Eng.* **2**(1).

[6] Raju N, Chepury A, Vinayaraj V S, Arkatkat S and Joshi G 2014 A Simulation study for improving the traffic flow efficiency at an oversaturated intersection coupled with BRT *Proc. 12th Conf. Transp. Plan. Implement. Methodol. Dev. Ctries. (TPMDC)* (Mumbai: India) pp41-47.

[7] Fieldmen O 2012 *The GEH Measure and Quality of Highway Assignment Models* (London: England)

Factors influencing household participation in solid waste management (Case study: Waste Bank Malang)

S Maryati[1,2], **N F Arifiani**[1], **A N S Humaira**[1], **H T Putri**[1]
[1]Research Cluster for Regional and City Infrastructure System, School of Architecture, Planning and Policy Development, Institut Teknologi Bandung, Jalan Ganesha 10, Bandung 40132, Indonesia

[2]E-mail: smaryati@pl.itb.ac.id

Abstract. Solid waste management is very important measure in order to reduce the amount of waste. One of solid waste management form in Indonesia is waste banks. This kind of solid waste management required high level of participation of the community. The objective of this study is to explore factors influencing household participation in waste banks. Waste bank in Malang City (WBM) was selected as case study. Questionnaires distribution and investigation in WBM were conducted to identify problems of participation. Quantitative analysis was used to analyze the data. The research reveals that education, income, and knowledge about WBM have relationship with participation in WBM.

1. Introduction

Solid waste management is an important environmental measure related to sustainable development. The amount of waste in Indonesia has increased year by year along with population growth, welfare and people's lifestyle [1]. One of the strategies implemented by the Ministry of Environment to reduce the amount of solid waste is establishment of Waste Banks in 250 cities by 2014 [2]. Based on this background, there are large opportunities to study about Waste Bank development and its prospect in Indonesia. There are various characteristics and community behavior in Waste Banks. However, studies related to Waste Bank in Indonesia are very limited, for instance, waste bank prospect, problems in developing waste bank and community empowerment in waste bank.

A waste bank in Malang City, East Java Province, Indonesia called *Bank Sampah Malang* or Waste Bank in Malang (WBM) was selected as location of study. The objectives of this research is to identify factors influencing participation in WBM. Finally, problems of BSM participation are identified and some recommendations to improve BSM participation are proposed.

It is expected that the result of research can be one strategy in Municipal Solid Waste Management (MSWM) and infrastructure provision. In particular, the result of this research can improve WBM management and contribute to promote Waste Bank system in other areas.

2. Overview of Solid Waste Management in Indonesia and Waste Bank in Malang (WBM)

Solid waste management in Indonesia is regulated nationally in Law of Republic of Indonesia No. 28 Year 2008. Based on the law, waste reduction can be done through reduce, reuse, and recycle measure. Waste management process according to that law consists of waste separation, waste collection, waste transportation and transfer, waste treatment, and disposal.

Waste separation and reduction is still new paradigm in Indonesia. Only limited households who have applied waste separation and waste reduction measure. On the other side waste generation increases year by year due to population growth. Lack of waste infrastructure become a problem because of the condition. Waste infrastructure is very limited compared to waste generation that has to be managed. Most people still apply old paradigm of solid waste management, that is collect-transport-and dispose.

One of government policy to reduce the amount of waste was waste bank concept. Waste bank is a concept in which people can get money by waste separation and then transfer the waste to the 'bank'. The bank then process the waste to get new good or just sell the waste to other industries. This process will generate money, and the participants in waste bank can get money from the bank. One of big and success waste bank in Indonesia is waste bank in Malang.

WBM was launched on July 2011[3]. There are several types of group member in BSM as shown in Table 1. NON includes waste dealers who are belonging to private firm. The number of kinds of recyclable materials to be collected by BSM are 69 and the main categories are plastics, paper, glasses, and metals. All members must separate their waste by themselves in accordance with the categories defined by BSM.

Recyclable materials discharged from COM and SCH are stored in a temporary storage site, called a WBM unit. Generally, WBM units are located near schools or in the residential area of COM. The administrators of COM and SCH are responsible for checking appropriate separation and record the savings report. The recyclable materials which has been stored in the WBM unit is transported to the WBM by WBM staffs based on the arranged schedule. Several COM make handicraft products using the stored recyclable materials. For example, plastic bags and photo frames are made from plastics waste and cardboard, respectively. These products are sold in the WBM unit, WBM office or at exhibitions.

IND and NON can bring their recyclable materials to the WBM office. IND can save their profits in their accounts. NON can only receive cash by way of compensation for their recyclable materials.

Table 1. Types of groups in WBM

Abbreviation	Explanation
COM	A group consists of min. 20 members and 3 administrators
SCH	A group consists of min. 40 members and 3 administrators in school
IND	A person do not belong to any group, but they register in BSM individually
NON	A person do not register as member in BSM. Only receive cash as compensation for their recyclable materials.
INS	The recyclable materials are generated from activities in institution
WAD	Private firm which buy recyclable materials from scavengers and bring it to BSM

The recyclable materials carried to WBM office are separated into 100 types from 69 kinds of types. The selling prices of plastics to factories are different depending on type and cleanliness. WBM shreds and washes plastics to gain higher profit. The recyclable materials collected by the WBM is sold to factories who produce recycled products [4].

3. Methods

3.1. Aspects and variables

Based on previous studies ([6],[7],[8],[9],[10]), there are several variables explored in this study. The variables are classified into personal attributes and knowledge. Personal attributes consist of Gender, Age, Family size, Occupation, Education Background, Monthly Income, Activity in Community Organization. Knowledge aspects consist of Knowledge about 3R and Information about WBM. In this study motivation to participate in WBM was also explored. Motivation explored are Environmental Awareness, Health Concern, Financial Motive, Follow Others, Other Reasons. Table 2 explains aspects and variables used in this study.

Table 2. Aspects and Variables

Aspect	Variables
Personal Attribute	Gender, Age, Family size, Occupation, Education Background, Monthly income, Activity in Community Organization
Knowledge	Knowledge about 3R, Information about WBM

3.2. Method of data collection and analysis

Data used in this study was primary data collected by distributing questionnaire. In this research, stratified random sampling technique was used to categorize respondents based on the characteristics [5]. In total, there were 381 questionnaires distributed. Respondents are divided into two types: (1) People join WBM (JOIN); and (2) People do not join WBM (NOT). The number of questionnaires collected were 211 for JOIN and 170 for NOT. The data from questionnaire were analyzed by descriptive analysis and association analysis.

4. Results and discussion

4.1. Results

The result of analysis shows that for personal attributes, only education background and monthly income show significant difference between JOIN and NOT groups. For knowledge aspect only information about WBM show significant difference between JOIN and NOT groups. The rest have no difference between group.

Education level of JOIN were lower than that of NOT. Sixty four percent of NOT have education from university and that ratio is higher than that of other groups. Mann-Whitney U test show significant (medium to large) difference of education background distribution between JOIN and NOT (probability = 0.000; effect size: -0.31333). There was a tendency that people with high education do not join WBM.

Although WBM have already conducted campaign in Malang City through many media, there were 51% of high income respondents do not know about WBM. Beside conducted campaign in other media, WBM officer have already conducted campaign in community organizations for all level of income, even though only one third of high income respondents experienced this campaign, lower than others income level.

Crosstab analysis for 3R knowledge shows that most of people from JOIN knows about 3R (Reduce, Reuse, and Recycle) concept. Most respondents know about this concept from campaign in communities, television, and their neighbors.

Crosstab analysis for WBM information shows that 64% of NOT group, do not know about WBM. It is an indication that the campaign from WBM is not enough and the improvement of method is necessary. Most people get information about this program from campaign in communities (46%), neighbors (17%), and friends (8%). Although campaign in community organization succeed, the announcement should be focused on those who not actively participate in local communities.

Chi Square analysis indicates that there was a strong difference of BSM information between JOIN and NOT. Information about WBM is an important factor which associated with participation in WBM.

Related to motivation to participate in WBM, 52% and 27% of respondents answered that environment and economic benefit respectively are their main motivation in participating in WBM. The percentage of invitation from other people, such as neighbors, family, or friend was 11%. Environment concern and economic benefit were chosen by half of respondents as their motivation participating in WBM. This result showed that financial and environmental concern are important aspects for participation in WBM. This result shows that not only financial benefit, but also environmental concern should be emphasized in WBM campaign.

In this study, lack of knowledge about WBM information are determined as reasons for not participate in WBM. Moreover, it is found that improvement of campaign method is required to increase WBM participation. Training of Trainers (ToT) for WBM unit administrators, city event participation, publication of success stories in mass media, should be conducted/improved. In cooperation with local government of Malang City, universities, community organizations, and mass media, campaign of WBM could reach all level of residents in Malang.

Participants in WBM are dominated by people from lower education background and income. It is actually related to the main reason participating in WBM, that is economic profit. People from lower education background and income use WBM as a mean for increasing their income. By participating in WBM, people can get extra money. People from higher education background and income, feel that the extra money from WBM is not very important, especially when it is compared to the effort to separate and manage the waste. People from JOIN group with higher education and income has motivation related to environmental concern.

Knowledge about WBM is important related to participation in WBM. People who join WBM knows it better than those who not join. If knowledge about WBM is compared to education background and income, it can be concluded that people from lower education and income have better knowledge about WBM. It is because they have enough time and willing to receive information about WBM.

5. Conclusions

Target group for participants in WBM can be focused on lower education and income group. It means that infrastructure and facilities for WBM can be developed with focus on the location of housing from lower education and income. People from lower education and income have motivation related to economic benefit in participating in WBM. In order to increase participants in WBM, campaign is very important. Campaign can be done by using media or community organisation.

Further investigation about cost and benefit aspect, human resources, social conflicts, and government policy related to BSM system are interested to be identified as a consideration to waste bank development in other areas.

6. References

[1] Ministry of Environment Republic of Indonesia 2008 Statistic of Solid Waste in Indonesia Year 2008 (Jakarta: Ministry of Environment Republic of Indonesia) (in-Indonesian)

[2] Ministry of Environment Republic of Indonesia 2011 Waste Bank Obtained 9 Million per Year Retrieved January 2013, from Ministry of Environment Republic of Indonesia Website: http://www.menlh.go.id/home/index.php?option=com_content&view=article&id=5147%3Abank-sampah-hasilkan-9-juta-per-bulan&catid=43%3Aberita&Itemid=73&lang=id (in-Indonesian)

[3] Ministry of Environment Republic of Indonesia 2012 Profile of Waste Bank in Indonesia 2012 (Jakarta: Ministry of Environment Republic of Indonesia) (in-Indonesian)

[4] Hidayat R 2012 Prospect of 3R Solid Waste Management Program through Waste Bank Management (Malang) (in-Indonesian)

[5] Purwanto E A and Sulistyastuti D R 2007 Quantitative Research Methods for Public Administration and Social Issues (Yogyakarta: Gava Media) (in-Indonesian)

[6] Aceti Associates of Arlington 2002 Recycling: Why People Participate; Why They Don't. (Massachusetts: Massachusetts Department of Environmental Protection)

[7] Kartini 2009 Factors Influencing Participation in Waste Bank and Impact of Waste Bank Establishment (Bogor: Institut Pertanian Bogor) (in-Indonesian)

[8] Permatasari D and Damanhuri E 2012 Effectiveness of Waste Bank as an Approach to Community Based Solid Waste Management (Bandung: Institut Teknologi Bandung) (in-Indonesian)

[9] Pradabphetrat P 2010 The Potential of Recycle Waste Bank Project in Reduction of Community Waste: A case of Sub-Sin Pattana Community, Bang Khun Thien, Bangkok (Thailand: Mahidol University)

[10] Singhirunnusorn W, Donlakorn K and Kaewhanin W 2012 Contextual Factors Influencing Household Recycling Behaviours: A Case of Waste Bank Project in Mahasarakham Municipality *Procedia - Social and Behavioral Sciences* **36** pp.688-697

A Computer Clone of Human Expert for Mobility Management Scheme (E-MMS): Step toward Green Transportation

Resdiansyah[1, 2], **R A O.K Rahmat**[3], **A Ismail**[3]
[1]Pembangunan Jaya Center for Urban Studies, Universitas Pembangunan Jaya, Cendrawasih Raya B7/P, Tangerang Selatan, Indonesia
[2] Civil Engineering Department, Universitas Pembangunan Jaya, Cendrawasih Raya B7/P, Tangerang Selatan, Indonesia
[3]Sustainable Urban Transportation Research Center, National University of Malaysia, Bangi, Selangor, Malaysia

Abstract. Green transportation refers to a sustainable transport that gives the least impact in terms of social and environmental but at the same time is able to supply energy sources globally that includes non-motorized transport strategies deployment to promote healthy lifestyles, also known as Mobility Management Scheme (MMS). As construction of road infrastructure cannot help solve the problem of congestion, past research has shown that MMS is an effective measure to mitigate congestion and to achieve green transportation. MMS consists of different strategies and policies that subdivided into categories according to how they are able to influence travel behaviour. Appropriate selection of mobility strategies will ensure its effectiveness in mitigating congestion problems. Nevertheless, determining appropriate strategies requires human expert and depends on a number of success factors. This research has successfully developed a computer clone system based on human expert, called E-MMS. The process of knowledge acquisition for MMS strategies and the next following process to selection of strategy has been encode in a knowledge-based system using a shell expert system. The newly developed computer cloning system was successfully verified, validated and evaluated (VV&E) by comparing the result output with the real transportation expert recommendation in which the findings suggested Introduction

1. Background

Demand for travel increases with the assumption that private cars are easy, comfortable, door to door services and judge by traveller as the only solution for mobility. In addition, supplies for infrastructure development such as roads are significantly increase and this is proportional to the increase in traffic volume. At one point, the supply will not be able to meet demand, due to certain factors such as budget problems, limited land or space. According to Frame, 2010 [2], It has become unacceptable that an uncontrolled traveling using private cars in urban areas cannot be accommodated. This is due to concerns about cost factors and the adverse impacts of traffic on local communities and the environment. If we let traffic grow to a level where there is a chronic and widespread congestion and often occurs on certain sections then it will economically become inefficient. On the other hand, although the construction of new roads to increase road capacity may reduce some of the effects of congestion but the benefits can only be matched by the growth of controlled traffic volume. Therefore, another alternative is to implement the Mobility Management Scheme (MMS), i.e. to ensure that people are still traveling but at the same time reducing the use of private cars [5]. In addition, the MMS application initiative will encourage people to reduce the number of trips they travel, travel frequently but use alternatives to driving alone, riding outside peak hours, and overall reducing the length of the trip they make. The main purpose of MMS is to minimize travel by private vehicle during peak hours and at the same time reduce the need for new road construction. On the other hand, MMS measures make driving alternatives more attractive, build a positive public attitude towards these alternatives, and provide information and incentives to encourage responsible travel behaviour.

1.1. Defining MMS

The Mobility Management Scheme (MMS) and Vehicle Miles of Travel [VMT] refers to a policy and program that can transform travel activities to achieve planning goals and improve the efficiency of a transportation system [5]. The size of an MMS consists of various types of interventions, especially in urban areas that aim at reducing transportation demand in the morning peak until afternoon peak hours or during a certain period.

Figure 1. MMS promote efficient use of street space.

2. How Computer Clone of Human Expert Can Help Us

In recent year, the MMS application guide manual has increasingly accepted as a tool for reducing traffic congestion during peak hours and reducing air pollution problems. A number of documents are available to guide people in implementing and marketing MMS programs, but few guidelines are offered regarding MMS program planning and how to select appropriate and efficient MMS strategies. Consequently, many MMS strategies are ineffective and developed with inefficient trial and error. Furthermore, the implementation of an MMS strategy requires an expert, as with their knowledge they can provide advice on the most appropriate MMS strategy based on location details. Unfortunately, very few expert people can be consider as expert, and are unlikely to be available at all times. These experts are not always there, nor do they always have time to consult all possible references, review existing data, etc. [3]. This will lead to time-consuming scheduling and interview scheduling, thus delaying a project. There is also the possibility that there may be no experts left behind after death, as most of them are in elderly age. Later on, it is a burden for Junior Engineer to refer to the book as part of their references to find solutions in implementing MMS [3]. On the other hand, Md Nazrul Islam [3] stated also in his research that the computer clone of human expert system has the potential to capture such expertise. The development of a computer clone of human expert system is designed to help employers, developers, junior transportation engineer, city council officer, property owners and managers, transportation management associations (TMAs), and MMS consultants. This a computer clone of human a computer clone of human expert system suggests a process for MMS development and implementation, and offers guidance and advice on the selection of effective MMS strategies. It will help to give specific description to choose a more effective MMS program in less time and with less frustration. It provides an overview of the "big picture" of MMS strategy, and then guides us identifying what strategies make sense for us, and in determining how much of those strategies we need to achieve a desired or required trip reduction.

In summary, development of a computer clone of human expert system for implementing suitable strategy in MMS in order to assist junior engineer and the policy maker in MMS would be a valuable tool. The development of a computer clone of human expert system requires both knowledge from transport and software engineers. Software engineers will develop a computer clone of human expert system based on civil engineer’s knowledge, sources from books and interview expert persons for implementations of MMS. Therefore, it is crucial to have knowledge from both fields in order to develop a computer clone of human a computer clone of human expert system as neither can do it alone.

3. A Computer Clone of Human Expert System Architecture

3.1. Knowledge Engineering

There are currently two ways of building a computer cloning of human expert. They can built from zero, or using development software known as "shell". Although there are different styles and methods of knowledge engineering, the approach was the same, knowledge eengineers (KE) interview and observe human expert or a group of expert people as domain and know what their intellectual think, and how they practice their knowledge. KE then translate that knowledge into computer-usable languages, and design inferencing machines, the structure of reasoning, which uses knowledge appropriately. KE will determines how to integrate the use of uncertain knowledge in the reasoning process, and what kind of explanation is useful for the end user. The next step is the inference engine and the facility to represent programmed knowledge, and domain knowledge will put into the program piece by piece. At this stage, the inference engine may be incorrect; the awkward form of knowledge representation for the kind of knowledge required for the task; and expert people can decide that piece of knowledge is not correct and finally discovered and modified as the computer cone of human expert system gradually gains competence.

In this project, the knowledge base of the computer clone of human expert system represents expertise in the domain of advice on MMS strategies. Expertise has been acquire and codified in the form of a production rules, an object entity in a particular hierarchy. In this project, knowledge base of computer clone of human expert system represents expertise in the domain of advice on MMS strategies. The expertise has been acquired and codified in the form of production rules, object entities in a specified hierarchy.

3.2. Knowledge Acquisition of E-MMS

The Victoria Transport Institute handbook and encyclopedia, Kuala Lumpur City Council Plan 2020 in MMS section, expert domain and practitioners of MMS in South East Asia accumulated massive knowledge about from its involvement and experience planning and devised of MMS practice and application in many countries in the world especially in SEA. Those parties involved in massive knowledge as stated above are primary knowledge source for this reserach.

3.3. Knowledge Abstraction and Representation of E-MMS

In a computer clone of human expert system, knowledge means the information that a computer system needs before a computer clone of human expert system can be intelligent. It needs to extract it first from the source of knowledge and then represented in a way that can be manipulate by the computer very efficiently. One is using rule-based representation. This method uses the conditions of IF and THEN statement actions to represent knowledge. When conditions in the IF section are met or adjusted, the rules will be eliminated and the actions determined by the THEN section are then triggered. Another representation method of knowledge is frame-based. It uses a network of classes and instances that are link by a relation and organize into a hierarchical structure. Figure 2 shows the relationship between classes and samples.

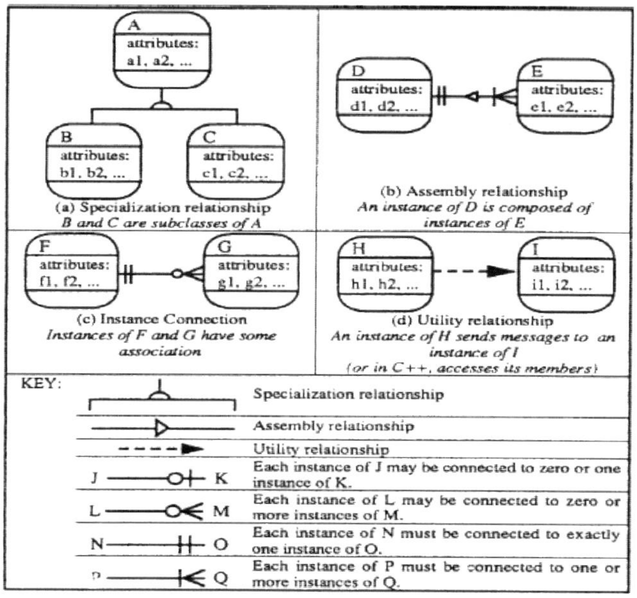

Figure 2. Class and instance relationship

Each node is represented by a concept that can be translated by attributes and/or values associated with that node. The topmost node represents the general concept and the node underneath represents a more specific concept. In this research, a shell-based software has been used. This software supports both methods of knowledge representation. However, the nature of the computer clone of human expert system in MMS makes this frame-based representation more efficient and broad-minded. In a shell software environment called Kappa-PC, the taxonomic process is following a common format as shown in Figure 3 below:

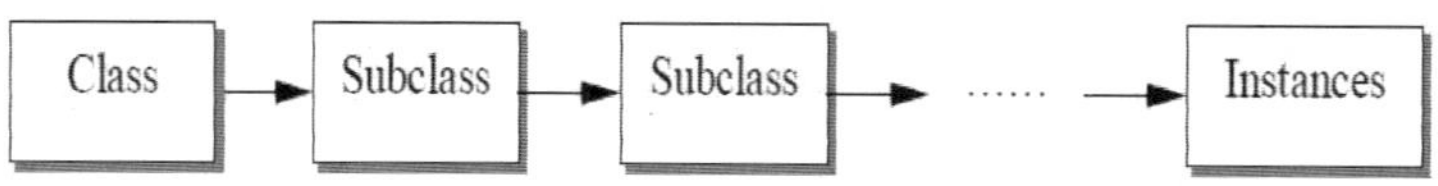

Figure 3. General format of taxonomic process.

In the system applications, the uppermost class was labeled as MMS, which has a broad meaning and includes all the strategies involved in MMS. Furthermore, it is group into nine (9) subclasses: congestion reduction, energy conservation, health and fitness, equity enhancement, livability strategies, parking solutions, rural communities, safety strategies, and affordability. All of these subclasses represent the main objectives in the MMS strategy application. Each of the nine (9) subclasses is divide into three (3) more detailed subclasses, which are part of the main categories of MMS namely better transportation options, incentives to use alternative modes and reduce driving, and land-use management. The final subclasses represent eight (8) geographical areas for each three (3) previous subclasses; large urban, urban, suburban, town, rural, commercial center, residential, and resort. Under each geographical area subclass, there are three (3) instances indicate the most suitable MMS strategies to be implemented according to its hierarchy. Details of the selection process suitable MMS strategies to be implemented in appendix.

3.4. Tools, Shells and Skeletons

There are many tools exist in developing a computer clone of human expert system. Three main categories are divided in the major tools used in practice:

- Programming language

- Programming environment
- An expert system shells

Selection of Kappa PC software is due to suitability in designer mode, which gives the knowledge engineer the possibility to code the knowledge base in a graphical (easily readable and editable) way that follows the latest trends in user interface. It lets the engineer decide what method to use for coding the knowledge, thereby addressing more complex problems. There is an opportunity to evaluate the shell system being built prior to the proposed advanced development, i.e., with access to a series of sample applications, and this has been recommended as a fairly effective tool for development. Experience throughout the study using the same shell system has shown that there is a high degree of confidence in the use of Kappa-PC [4], and this is a useful tool for developing an understanding of expert systems. An important issue was the prototype development process that utilized Kappa-PC as platform in a computer clone of human expert system shell (as the building-tool). Expert system shell was preferred since it offered the advantage of ease of manipulation, increased productivity and a true windowing capability to run on personal computers. Building a computer clone of human expert system shell offers significant advantages. A system can be built to perform a unique task by entering into a shell all the necessary knowledge about a task domain. The inference engine that applies the knowledge to the task at hand is built into the shell. If the program is not very complicated and if an expert has had some training in the use of a shell, the expert can enter the knowledge himself.

3.5. Domain Acquaintance

It is important for a knowledge engineer (KE) to gain a good understanding of the domain to enhance the building of the prototype with respect to its feasibility and evaluation [1]. The first and foremost step for the knowledge engineer is to gain knowledge engineering skill.

3.6. Domain Expert

The domain or area expert is an articulate, knowledgeable person with a reputation for producing good solutions to problems in a particular field. The expert uses tricks and shortcuts to make the search for a solution efficiently, and a computer clone of human expert system models these problem-solving strategies. A person's ability to serve as a domain expert will depend on characteristics such as the ability to explain important knowledge or heuristics, be introspective, be patient, and communicate effectively. There are broad categories of expertise sources to be identified and selected from: textbooks and manuals written by experts and related professional institutions; experts involved in the domain; and research papers from journals and conferences. The advantages of multiple sources of expertise are as follows: bias towards a single view will be avoided; conflicting views manifested by several different experts provide a genuine picture of uncertainty of the subject matter or indicate that further research is required; the knowledge engineer will not be bound or handicapped by availability constraints of any one source of expertise; and completeness of knowledge will be readily achieved.

3.7. Source of the Knowledge Base

The main knowledge was obtained through communication with the transportation experts and engineers who involved directly with the MMS strategy and implementation. The domain experts were selected from experienced MMS planner from several countries in South East Asia. More than 30 transportation experts involved in this project. These experts were chosen from individuals who were currently practicing MMS in their city and some transportation expert from university who had involved in transportation field for many years. On the other hand, the selection of domain experts also critically depending on the expert's availability (time), possess of good communication skills, willingness to participate, and acknowledged as expert by the community or by their peers. Due to time constrains, most of the interview and consultations with the domain experts were done through questionnaire and if knowledge engineer successful to interview the expert, it is done informally and not recorded by tape due to certain reason that expert's not comfortable if the interview recorded by tape. The questionnaire was designed in such a way to covered all nine main objectives in MMS which were obtained previously through different sources from text books and manuals

IOP Conf. Series: Earth and Environmental Science **124** (2018) 012016 doi:10.1088/1755-1315/124/1/012016

3.8. Object Representation of E-MMS

The object browser window in Figure 4 represents the top-level hierarchical in the object model developed for the E-MMS domain. The class E-MMS represents the overall prototype that is comprised of nine major subclasses.

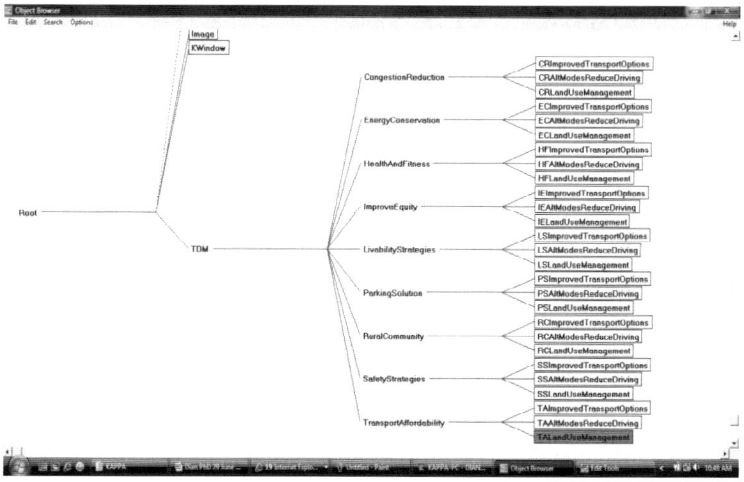

Figure 4. Top-level object hierarchy of the E-MMS.

3.9. User Interface Design and Example of E-MMS Module

Kappa-PC provides powerful tools for creating graphic user interfaces. In addition to session windows, predefined images, custom menus, and dialog windows, it can import Visual Basic (VBX) controls to meet knowledge engineer's specific requirements. The graphic user interfaces started with general information interface which is the first window invoked after the E-MMS is launched, as shown in Figure 5. It contains information about the E-MMS title, developer, version series number, and development date.

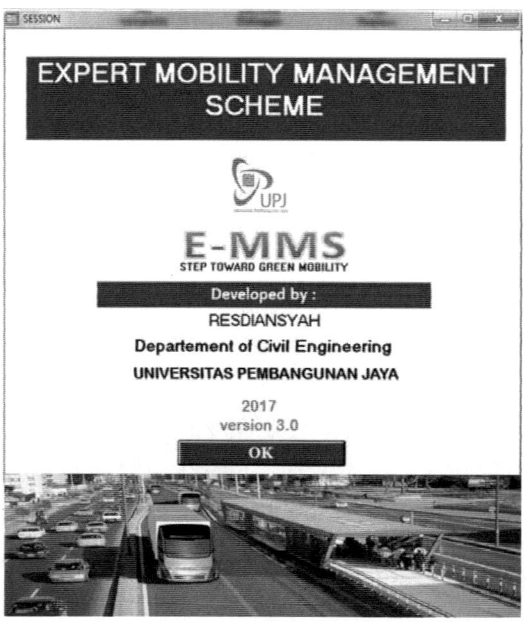

Figure 5. General information interface.

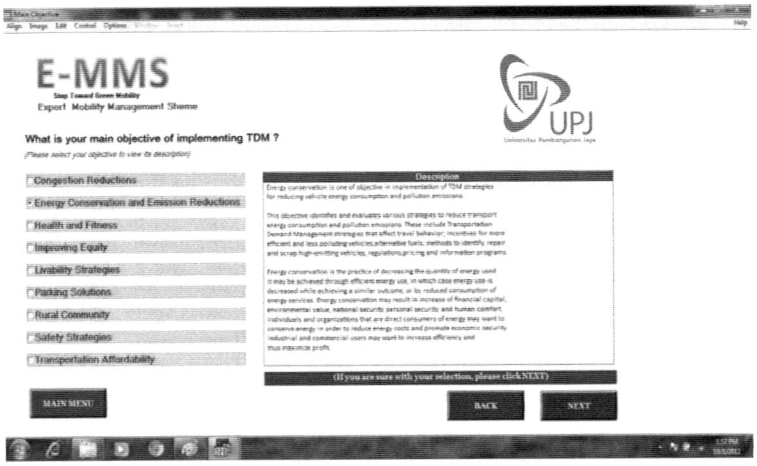

Figure 6. Screen snapshots example of one module in E-MMS

Furthermore, when all instruction in E-MMS has been applied, finally system will give advice strategy summary (Fig. 7) which show the travel impact, benefit and equity if the strategy chosen will be implemented. Based on the travel impact summary, user can easily to identified how effective reduction of trip or total traffic which is stated in the summary using the rating scale; rating from 3 (very beneficial) to -3 (very harmful); 0 indicates no impact or mixed impact on the objective if user implement this strategy. If the user can confirm this strategy is the most suitable strategy, then it is necessary to clicking PRINT button to viewing window spawns a notepad text editor with analysis result (rating scale).

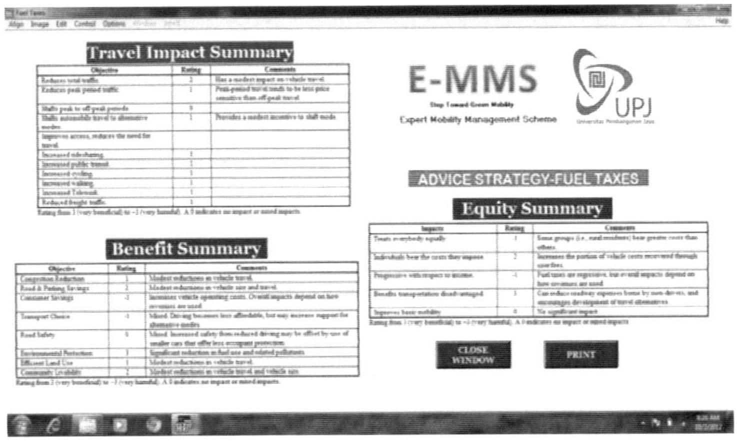

Figure 7. Impact summary interfaces.

4. Verification, Validation and Evaluation (VVE)

The prototype testing or validation is the process of evaluating the performance and utility of the program and revising it as necessary. Testing the prototype also involves looking at the ease of handling the user interface and understanding the explanatory text [1]. The verification starts with learning the behavior of E-MMS under different sub-module in order to studied the consistency and stability of the system. In the E-MMS, there are nine different sub-modules which are dealing with the same specific aims to achieve main goal or objective of MMS strategies. The first scenario was the application of first sub-module under Congestion Reduction and applied to specific aims of Improve Transport Option. When the input was selected then the system will have presented the description of Improved Transport Option in the dialog box. This description should appear as same information when input was selected for second sub-module for Energy Conservation with same specific aims (Improved Transport Option). The results were identical between the different input scenarios. When testing continued to the other seven sub-modules, the E-MMSA still maintain the consistency. Hence, this reflects high stability and reliability of the developed system. The validation result shows that more than 60 %, or exactly 69 % matched between E-MMS decision and evaluators who are also transport specialist. The result of 69% has been exceeded the level of performance which is expected will be only 50-60 % matched since the strategy is very subjective to evaluate and every single expert either domain expert or evaluator who involved in validation process has different experiences in implementing of MMS strategies. The percentage accuracy in this system is the level of performance that is acceptable to the users is called the acceptable range of performance, and should be specified at some stage of development [1]. It can be specified by a third party, a government agency or the project sponsor. In this case, an acceptable range performance had been specified by a government (DBKL) with range of 50-60% during knowledge acquisition process. This is indicating that the possibility of using the E-MMS as expert system is encouraging. Thus, we can say that the system developed has achieved the purpose that it was designed for. Generally, the flow of the consultation session or advice session between user and the E-MMS system was satisfactory. The user was able to learn every single description in MMS strategy and able to obtain recommendations given by the system. The system also allowed the user to view the impact of result before make decision on the appropriate MMS strategies to be implemented in particular situation based on the input selected. The E-MMS interface was built to accommodate all parameters and variables as stated above and the interface built was considered to be adequate in its design and functioning.

5. Conclusion and Recommendations

5.1. Conclusions

The successful development of an expert system for implementations of MMS demonstrates that the application of a computer clone of human expert system in this domain is promising. Frame-based knowledge representation method in conjunction with the power of object-oriented programming enables the knowledge to be abstracted and represented for the computer to manipulate efficiently and for the knowledge engineer to update the knowledge base with great flexibility. Since the knowledge in this domain is dynamically growing, the E-MMS is by no means complete at this stage.

5.2. Recommendations

The knowledge in MMS strategies are constantly growing and their effectiveness relies on the country itself. The developed E-MMS must also reflect the latest information in this domain. Therefore, it is necessary to periodically assess the new knowledge and loaded to the knowledge base of E-MMS. A newer version of E-MMS may consist of new and updated information and the implementations of MMS strategies and must be reflecting the condition and situation of study area.

6. References

[1] Basri H 2000 An Expert System for Landfill Leachate Management (J. Environmental Technology vol 21) (Elsevier Science Ltd, London) p 157-166

[2] Frame G 2010 Traffic Management and Transport Demand Management (Distance Learning Course in Urban Planning) chapter 4 pp 1-19.

[3] Islam N 2005 An Expert System for Mix Design of High Performance Concrete (J. of Advances in Engineering Software vol 36) (Oxford: Elsevier) p 325–337

[4] IntelliCorp Inc 1997 Kappa-PC Ver. 2.4 User Manual (California:IntelliCorp Inc).

[5] Litman T 2013 Are Vehicle Travel Reduction Targets Justified? Evaluating Mobility Management Policy Objectives Such As Targets to Reduce VMT And Increase Use Of Alternative Modes (online) Victoria Transport Policy Institute pp 1-34

Rapid urbanization and the need for sustainable transportation policies in Jakarta

D Rukmana[1]
[1] Urban Studies and Planning Program
Savannah State University, Savannah, GA, United States

Email: rukmanad@savannahstate.edu

Abstract. Not only is Jakarta the largest metropolitan area in Southeast Asia, it is the also one of the most dynamic, though beset with most of the urban problems experienced in twenty-first century Southeast Asia. Batavia, colonial capital of the Netherland Indies in the first half of the 20th century was a small urban area of approximately 150,000 residents. In the second half, Batavia became Jakarta, the 28 million megacity capital of independent Indonesia. Among many urban problems, one major problem plagued Jakarta in the last two decades is traffic congestions. This paper discusses the extent to which rapid urbanization in Jakarta has contributed to the need for sustainable transportation policies in Jakarta. The development of MRT could be viable solutions to alleviate the acute traffic jams in Jakarta. Jakarta will need to implement other innovative sustainable transportation policies including promoting active live through more walking and bicycling, carpool matching services, shuttle services, telecommuting and downzoning in downtown areas.

1. Introduction

Jakarta is the capital of Indonesia and the largest metropolitan area in Southeast Asia with tremendous population growth, land use change and new town and industrial estate development. The overall population of the Jakarta region grew in the 20th Century, from about 150,000 in 1900 to about 30 million in 2014 (Silver 2007; Firman and Fahmi 2017). This paper discusses urbanization and suburbanization in the megacity of Jakarta and analyzes the extent to which rapid urbanization in Jakarta has contributed to the need for sustainable transportation policies in Jakarta. The development and expansion of Mass Rapid Transit are documented and the use of private vehicles and the development of elevated toll roads and bike lanes also critically analyzed.

1.1. Population Growth of the Metropolitan Region of Jakarta

Jakarta has been the capital of Indonesia since the Dutch colonial era. The population of Jakarta in 1900 was about 115,000. In the first nationwide census of the Dutch colonial administration (1930), Jakarta's population increased to 409,475. In the next ten years, the population increased to 544,823 with an annual growth rate of 3.30%. After Independence, Jakarta increased by nearly three times to 1.43 million by 1950. It increased to 2.91 million in 1960 and 4.47 million in 1970. The annual growth rates of Jakarta's population are 10.35% and 5.36% (1950-1960 and 1960-1970 respectively) (Firman 1997 and Cox 2011).

Table 1 shows the population of the metropolitan region of Jakarta including Jakarta, the inner and outer peripheries of Jakarta, from 1980 to 2010. The Megacity of Jakarta increased from 11.91 million in 1980, 17.14 million in 1990, and 20.63 million in 2000 to 28.01 million in 2010. The megacity in 2010 was 11.79 percent of Indonesia's total population but this population resides in less than 0.3 percent of Indonesia's total area. The proportions of Jabodetabek's population to the total population of Indonesia have steadily increased from 8.07%, 9.56%, to 10.0% (in 1980, 1990, and 2000 respectively).

Table 1. Population of the Metropolitan Region of Jakarta in 1980-2010 (in millions)

Area	1980	1990	2000	2010
Core	6.50	8.26	8.39	9.60
Jakarta	6.50	8.26	8.39	9.60
Inner peripheries	n.a	n.a	4.93	7.22
City of Tangerang	n.a	n.a	1.33	1.80
City of South Tangerang	n.a	n.a	0.80	1.29
City of Depok	n.a	n.a	1.14	1.75
City of Bekasi	n.a	n.a	1.66	2.38
Outer peripheries	5.41	8.88	7.31	11.20
City of Bogor	0.25	0.27	0.75	0.95
Tangerang Regency	1.53	2.77	2.02	2.84
Bekasi Regency	1.14	2.10	1.62	2.63
Bogor Regency	2.49	3.74	2.92	4.78
Megacity of Jakarta	11.91	17.14	20.63	28.02

Sources: Rukmana (2014)

1.2. Transformation of Jakarta

The modern city of Jakarta was initiated by President Soekarno's strong vision to build Jakarta into the greatest city possible (Cybriwsky and Ford, 2001). He gave Jakarta, Monas – his most symbolic new structure the 132 m high national monument, spacious new government buildings, department stores, shopping plazas, hotels, the sport facilities of Senayan that were used for the 1962 Asian Games, the biggest and most glorious mosque of Istiqlal, new parliament buildings and the waterfront recreation area at Ancol.

Such constructions continued under the New Order regime that began in 1967. Under this regime, Indonesia enjoyed steady economic growth, along with a reduction in the percentage of the population living under the poverty line. Jakarta grew rapidly during this period of the New Order regime. During the thirty-two years of the New Order regime, Jakarta changed considerably. A generally rapid economic growth during this period allowed Jakarta to expand its modern constructions and develop into a modern city. Hundreds of new office towers, hotels and high-rise condominiums were built in many parts of the city.

The massive development on the outskirts of the megacity of Jakarta resulted from a series of deregulation and de-bureaucratization measures enacted by the Suharto government in the 1980s (Winarso and Firman 2002, p. 488). The subsidized housing finance program and municipal permit system for land development also contributed to policies that have most benefited some developers strongly linked to the New Order regime (Leaf, 1994). Winarso and Firman (2002) revealed almost all large developers were well connected to the President Suharto's family and inner circle including his daughters, sons, brother, in-laws and close friends. The connection to the Suharto family and inner circle became significant; closeness to the first family helped the large developers expand their business. Interlinking also occurred among the large developers through cross-shareholding, shared directorships and joint ventures; process which turned potential competitors into collaborators and created oligopolistic types of land and housing markets.

Continuing with the suburbanization; this was also caused by the development of three highways stretching from Jakarta to the peripheries - the Jagorawi toll road, the Jakarta-Cikampek toll road, and the Jakarta-Merak toll road (Henderson and Kuncoro 1996). The development of private industrial parks in the peripheries naturally followed the development of these highways (Hudalah et al 2013). Private industrial parks in the peripheries range from 50 to 1,800 hectares and on average the size is about 500 hectares (Hudalah et al 2013); major industrial centers are located in Cikupa-Balaraja of Tangerang Regency and Cikarang of Bekasi Regency. The industrial center of Cikarang with a total industrial land area of nearly 6,000 hectares is the largest planned industrial center in Southeast Asia (Hudalah and Firman 2011).

2, Jakarta's Transportation Problems

The urbanization and suburbanization in Jakarta are strongly associated with the traffic congestion in Jakarta. Jakarta is estimated to lose US$3 billion a year because of traffic congestion which can't be separated from the high growth rate of vehicle ownership (9 to 11 percent per year), unsupported by road development (less than 1 percent a year). (Rukmana 2010)

Motorcycles are ubiquitous and can be acquired with a down payment of as little as $30. The number of registered motorcycles in Jakarta grew exponentially from 2000 to 2010 as seen in Figure 1. The trend of the motorcycles growth will continue until a more sustainable transportation policy is implemented.

People who live in the outskirts of Jakarta can save as much as 30% of their transportation costs using motorcycles to work rather than public transport (HCC 2017). The daily jams in Jakarta are getting worse; the peripheries are a "bedroom suburb" for the daily commuters of Jakarta, the center of government and corporate offices, commercial and entertainment enterprises. The economy of Jakarta dominates its peripheral areas. In the daytime, the total population in Jakarta is much more than its population in the nighttime; the number of daily commuters in Jakarta is estimated 5.4 million (Rukmana 2014),

Figure 1.
Number of Registered Vehicles in Jakarta in 2000 - 2010

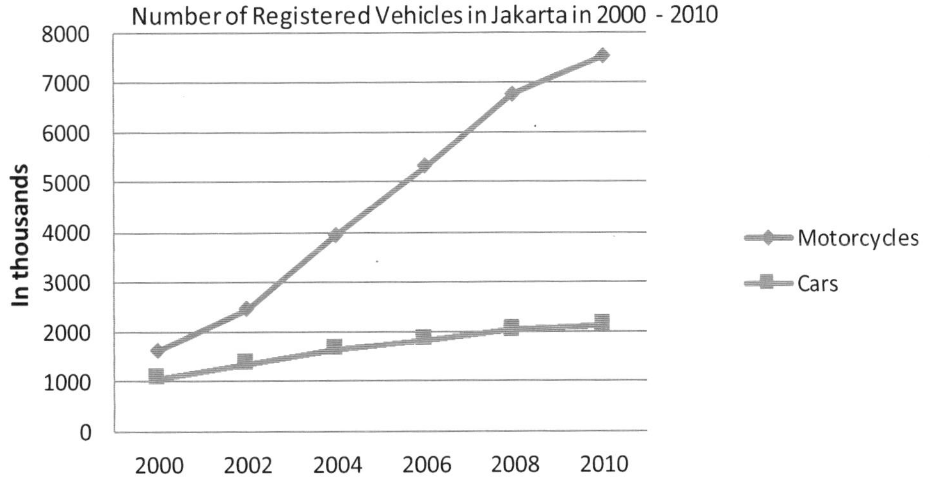

Source: JUTPI (2011) and BPS DKI 2011

The economy of Jakarta dominates its peripheral areas. In the daytime, the total population in Jakarta is much more than its population in the nighttime; the number of daily commuters in Jakarta is estimated 5.4 million. The level of services of public transportation in Jakarta is not reliable and accessible. People who live in the outskirts of Jakarta can save as much as 30% of their transportation costs using motorcycles to work rather than public transport (HCC 2017)

Commuters from the peripheries primarily used three highways including the Jagorawi toll road connecting Jakarta and the southern peripheries, the Jakarta-Cikampek toll road connecting Jakarta and the eastern peripheries and the Jakarta-Merak toll road connecting Jakarta and the western peripheries. Most commuters go to Jakarta to work in government and corporate offices, study in universities, receive high quality medical attention in the hospitals, and/or go for entertainment and cultural activities. The current public transportation systems have not been able to alleviate the acute traffic congestion; it is likely that Jakarta needs a Mass Rapid Transit (MRT) or also popularly known as Metro, in order to address this problem. The construction of the MRT project in Jakarta began on October 10, 2013. The first MRT tract will connect Lebak Bulus, South Jakarta and the Hotel Indonesia traffic circle with six underground stations, seven elevated stations and a capacity of 173,000 passengers per day (Rukmana 2014). By June 30, 2017, the completion of the MRT first tract was nearly 75 percent. The Jakarta city administration expected to launch the service of MRT to the public for trial purposes in August 2018.

Most metropolitan areas in the world with the population of over 10 million have operated metros for years. New York City opened the first underground line of its subway in 1904 and since then the subway has been the backbone of New York City transportation system. Two major cities in Japan, Tokyo and Osaka built their metros in 1927 and 1933 respectively. The Tokyo Metro is the world's most extensive rapid transit system with more than eight million passengers daily. The second largest city in the world, Mexico City, has had a metro since 1969 and now the Mexico City Metro is the second largest metro system in North America after the New York City subway. Two major cities in China, Beijing and Shanghai opened their metro systems in 1971 and 1995 respectively. Major cities in Southeast Asia with smaller populations than Jakarta have also had their metro systems for years, including Manila (1984), Singapore (1987), Kuala Lumpur (1995) and Bangkok (2004).

3. The Need for Sustainable Transportation Policies in Jakarta

Rapid urbanization in Jakarta has contributed to the need for sustainable transportation policies in Jakarta. The growth of registered vehicles in response to the growth of population in Jakarta and its peripheral areas has caused acute traffic congestions in Jakarta. This section will offer several ideas of sustainable transportation policy including reducing private vehicle uses, promoting bike lanes, and the development of MRT.

3.1. Reducing private vehicle uses

There are several possible solutions to eradicate traffic congestion problems and one of them is the reduction of private vehicle uses. The New York Times reported a suburb town without cars in Germany. Streets in this upscale town are completely car-free except the main thoroughfare and a few streets on edge of the town. The residents of this town are still allowed to own cars, but parking is relegated to two large garages at the edge of the development.

The Vauban town, is located on the outskirt of Freiburg, near the French and Swiss borders and home to 5,500 residents. The residents are heavily dependent on the tram to downtown Freiburg and many of them take to car-sharing when longer excursions are needed. Seventy percent of Vauban's families have no cars. They do a lot of walking and biking to shops, banks, restaurants, schools and other destinations that are interspersed among homes. The town is long and relatively narrow and provides an easy walking access to the tram for every home.

Creating places with more compact design, more accessible to public transportation and less driving is the envision of urban planners in the 21st century. The Vauban town is an exemplar of the 21st century urban design in response to the threats of greenhouse gas emission and global warming and the dwindling oil supply. The Vauban's urban design is the extension of the New Urbanism. The New Urbanism is a school of urban design arose in the U.S. in the early 1980s. This school of urban design promotes several key principles including walkability and connectivity, mixed land uses, and high density. There have been many the New Urbanist towns in several countries, but cars still fill the streets of these towns.

The Vauban town provides an example of the possibility of creating city without cars. The walkable and mixed-land-uses urban design, easy access to public transportation and excellent public transportation system as demonstrated in the Vauban town are the components for creating city without cars.

Cars are still a luxury item for many Indonesian families. Many urban residents, particularly those live in kampung kota, do not own cars and are used to living without cars. Streets (gang) in Indonesia's kampung kota are too narrow for cars and the residents are used to walking and biking to their destinations. Kampung kotas are located in the center of urban areas and relatively accessible to public transportations. In reference to the New Urbanism concept, the Indonesia's kampung kota has implemented the principles of walkability and high density.

Indonesian planners need to appreciate the existence of kampung kota in terms of lacking driving needs. Kampung kota residents will be less likely to have a demand for cars when their neighborhoods are accessible to public transportations and the streets in their neighborhoods remain narrow. Kampung kota residents need to remain lack of driving needs for reducing the car ownership rate in urban areas including Jakarta. For new developments in suburb areas, Indonesian planners can emulate the success of the Vauban town. Driving needs are profoundly affected by the urban design and the high access to public transportation. It makes sense to envision and is not all impossible to create a city without cars.

3.2. Promoting Bike Lanes

Many metropolitans in the world have developed dedicated bicycles lanes for years. Cities in developed countries, particularly in Europe, have integrated bicycle lanes into their transportation network systems. Those cities such as Amsterdam, Paris, Berlin, Copenhagen and Barcelona have been developed as bike-friendly cities. Safe and extensive bike route networks, promotion of pro-cyclist policies, and a bike culture have taken places in those cities. Cyclists in those cities are not second class residents and can safely ride their bicycles as the main mode for their daily commute to their workplaces. Copenhagen is an example of European bike-friendly city where about a third of workforce in this city commute to the office by bike.

The first Jakarta's dedicated bike lane stretching from Ayodia Park to Blok M was inaugurated in May 2011. It should be considered as a breakthrough in solutions for acute traffic congestions in Jakarta. The development of dedicated bicycle lanes is a good move from the Jakarta administration for promoting the use of bicycle as an alternative transportation mode. If the Jakarta city administration could encourage more motorists to shift to using bicycle to work, the city's chronic traffic woes could be eventually reduced.

The first dedicated bike lane in Jakarta is only a small step in developing Jakarta as a bike-friendly city. There are many challenges for Jakarta to be a bike-friendly city. The Jakarta city administration needs to have a strong commitment to build more dedicated bike lanes and integrate them with the city transportation network system. Dedicated bike lanes should be part of the city transportation network system and designed to accommodate the need of residents' mobility in the city. It is very essential to connect dedicated bike lanes with mass transportations including the Mass Rapid Transit (MRT).

It's not easy to build more dedicated bike lanes if the Jakarta city administration still focuses on building more elevated inner-city toll roads as the solution of addressing the chronic traffic woes in Jakarta. It is also important to note that the first dedicated bike lane was not initiated by the Jakarta city administration but the Indonesian Bicycle Community (Komite Sepeda Indonesia) that donated as much as 500 million rupiahs to build the bike lane. The Jakarta city administration needs to change the mindset of the possible solution for the chronic traffic congestion in the city. The solution is not building more roads, but reducing the use of cars through improving and expanding the use of mass transportations and bicycles.

Another big challenge for bike lanes in Jakarta is the lack of law enforcement. The Jakarta city administration should strictly enforce the dedicated bike lanes for cyclists. The dedicated bike lane cannot be used as parking spots and a lane for motorcyclists. A few days after the inauguration of the bike lane stretching from Ayodia Park to Blok M, the lane was overwhelmed by private cars, pubic minivans and three-wheeled vehicles bajaj. A number of private cars were also parking in the lane (The Jakarta Post, 27 May 2011). Without strict law enforcement, the dedicated bike lane will not be an effective way to reduce the Jakarta's traffic woes and will only be a failed initiative.

Despite the challenges for bike lanes in Jakarta, the inauguration of the Jakarta's first bike lane should be seen as a promising way of alleviating the acute traffic problems in Jakarta. I hope that the inauguration of the Jakarta's first bike lane could be the milestone for the Jakarta city administration in changing the mindset of how to address the chronic traffic problems in Jakarta. It is not building more toll roads but reducing the use of cars through encouraging more motorists to shift to cyclists or mass transportation riders.

3.3. Discouraging Elevated Roads

The development of new roads will never catch up to the growth rate of vehicle ownership. A new highway or a widened road only alleviates traffic congestion for a short period of time. After a few years, any new or widened highway fills with traffic that would not have existed if the highway had not been built, a phenomenon called induced demand. Because of induced demand, neither building new roads nor widening existing roads are viable long-term solutions to traffic congestion.

The new roads will also undermine the efforts to develop a mass transportation system in Jakarta. The main idea of developing a mass transportation system, including the TransJakarta busway and the monorail and Mass Rapid Transit projects, is to reduce the number of motorists and motorcyclists on Jakarta's streets. Drivers would be expected to use the mass transportation and reduce traffic, but new roads would only attract more motorists.

Not only would elevated roads stimulate induced demand and thus worsen traffic congestion, they could also jeopardize the livability of neighborhoods along them. In many cities in other countries, such as Seoul, New Orleans, San Francisco and New York City, elevated freeways have negatively affected livability. At the same time, in many developed countries, we have seen a shift in urban planning from enhancing mobility toward promoting livability.

The Cheonggyecheon freeway was completed in 1977 and was seen a as a symbol of modernization and industrialization in South Korea after the war with the North. This elevated freeway was built above a 5.8-kilometer stream flowing through downtown Seoul. By 2000, the area was considered the most crowded and noisy part of the city and became an eyesore for residents.

In July 2003, the then-mayor of Seoul and the current president of South Korea, Lee Myung-bak, launched a project to tear down the Cheonggyecheon freeway and revitalize the surrounding area. During the demolition process, the city administration developed public transportation systems, including Bus Rapid Transit lines. Today, the Cheonggyecheon area has been revitalized and is one of Seoul's main tourist areas.

In 1973, New York City's West Side elevated highway collapsed and was never repaired but replaced by a surface boulevard of West Avenue. Similarly, two elevated freeways in San Francisco, Embarcadero and Central Freeways, were badly damaged by the Loma Prieta earthquake of 1989. The San Francisco city administration decided not to rebuild the elevated freeways, but replaced them with surface boulevards. The conversion of elevated freeways in both New York City and San Francisco did not cause traffic havocs. The traffic switched to the boulevards, nearby street or mass transit (James and Norquist 2010). Furthermore, a team of researchers from the UC Berkeley (Cervero, Kang, and Shively 2009) found that the conversion of elevated Embarcadero and Central Freeways with boulevard has stimulated reinvestment in the neighborhoods along the freeways without seriously sacrificing transportation performance. More recently, the residents of New Orleans have decided not to rebuild the damaged elevated expressway caused by the Hurricane Katrina, but replace it with an oak-lined boulevard (James and Norquist 2010).

The conversion of elevated freeways to surface boulevards in Seoul, New York City, San Francisco or New Orleans is evidence of a paradigm shift from a focus on expediting the movement of automobile to a focus on increasing the livability of neighborhoods. The livability of neighborhoods should be prioritized over the increase of mobility. Jakarta needs to learn from what has happened in Seoul, New Orleans, San Francisco or New York City regarding the elevated freeways. Not only is the proposed six elevated toll road projects the solution for the traffic congestion in Jakarta, but also they could cause the decline of livability of neighborhoods along the elevated toll roads. The Jakarta city administration should revisit their decision to build the new elevated toll roads and instead they should focus their efforts on building mass transportation systems in alleviating transportation problems in Jakarta.

3.4. The Development of Mass Rapid Transit

In order to address traffic congestion, two flagship projects are underway including the development of Cilamaya Seaport and the Mass Rapid Transit (MRT) project. The Cilamaya Seaport is located outside of the Jabodetabek, but is connected with the Jakarta-Cikampek toll road. This planned seaport is located in Karawang Regency. All shipping activities of industrial parks in the megacity are currently using the Tanjung Priok port in Jakarta. The Cilamaya Seaport is designed to mitigate further traffic congestion caused by heavy traffic flowing from the eastern parts of the megacity to the Tanjung Priok port. This will be a new transportation hub for the megacity's industrial parks. The MRT project would become the most expensive public projects in Jakarta's history, but it is the answer to address acute traffic congestions in Jakarta.

For at least 20 years, the proposed MRT has been under discussion by the Jakarta administration and the government of Indonesia. Activists and non-governmental watchdogs have seen the MRT proposal as a possible bonanza for corrupt politicians and contractors (Economist, 4 February 2010). Eventually, the government secured a $1.6 billion loan agreement with the Japanese International Cooperation Agency (JICA) in 2009 for funding. Vice President Boediono also asked the JICA to expedite the design and construction of MRT project; the first tract of the MRT project was to be completed in 2016 (The Jakarta Post, 20 October 2010). The construction of the MRT project began on October 10, 2013. Governor Joko Widodo launched the flagship project in a groundbreaking ceremony at Dukuh Atas, Central Jakarta (The Jakarta Globe, October 11, 2013).

One MRT train will consist of six cars and be able to transport a maximum of 1,200 passengers per trip. The MRT Jakarta will operate 16 trains and transport 1.5 million passengers a day. The first MRT tract will connect Lebak Bulus, South Jakarta and the Hotel Indonesia traffic circle with six underground stations, seven elevated stations and a capacity of 173,000 passengers per day.

4. Conclusion

The growth of registered vehicles in response to the rapid urbanization has caused acute traffic congestions in Jakarta. The Jakarta administration needs to find sustainable transportation policies to address the transportation problems in Jakarta. The development of MRT could be viable solutions to alleviate the acute traffic jams in Jakarta. The main idea behind developing a mass transportation system, including the TransJakarta busway and the monorail and Mass Rapid Transit projects, is to reduce the number of motorists and motorcyclists on Jakarta's streets. Drivers would be expected to use the mass transportation and reduce traffic, whereas new roads only attract more motorists. Jakarta needs to discourage the development of elevated roads that will stimulate induced demand and thus worsen traffic congestion. Elevated roads will jeopardize the livability of neighborhoods along them.

In addition, Jakarta will need to promote bike lanes and reduce the use of private vehicle. Jakarta needs to encourage the development and usage of the smartphone apps on its two-way ability to locate, coordinate and orchestrate both passengers and vehicles and encourage more biking and walking for its residents. Jakarta will also need to implement other innovative sustainable transportation policies including carpool matching services, shuttle services, telecommuting and downzoning and better parking management in downtown areas.

References:

[1] BPS DKI Jakarta. 2011. *Statistik Daerah Provinsi DKI Jakarta*. Jakarta: BPS DKI Jakarta
[2] Cervero, R., Kang, J., and Shively, K. (2009). From elevated freeways to surface boulevards: neighborhood and housing price impacts in San Francisco. *Journal of Urbanism* 2(1): 31-50
[3] Cox, Wendell. (2011). *The evolving urban form: Jakarta (Jabotabek)*. http://www.newgeography.com/content/002255-the-evolving-urban-form-jakarta-jabotabek
[4] Cybriwsky, Roman and Ford, Larry R. (2001). City profile: Jakarta. *Cities* 18(3): 199-210.

IOP Conf. Series: Earth and Environmental Science **124** (2018) 012017 doi:10.1088/1755-1315/124/1/012017

[5] Economist, 4 February 2010
[6] Firman, Tommy., & Fahmi, F. Z. (2017). The Privatization of Metropolitan Jakarta's (Jabodetabek) Urban Fringes: The Early Stages of "Post-Suburbanization" in Indonesia. *Journal of the American Planning Association*, 83(1), 68-79.
[7] Firman, Tommy. (1997). Land conversion and urban development in the Northern Region of West Java, Indonesia. *Urban Studies* 34(7): 1027-1046
[8] Human Cities Coalition (HCC). (2017). *Jakarta Urban Challenges Overview.* https://www.humancities.co/2017/01/jakarta-urban-challenges-overview/
[9] Henderson, V., and Kuncoro, A. (1996). The dynamics of Jabotabek development. *Bulletin of Indonesian Economic Studies*, 32(1), 71-95.
[10] Hudalah, Delik and Tommy Firman. (2011). Beyond property: Industrial estates and post suburban transformation in Jakarta Metropolitan Region. *Cities* 29: 40-48
[11] Hudalah, Delik, Dimitra Viantari, Tommy Firman, and Johan Woltjer. (2013). Industrial land development and manufacturing deconcentration in Greater Jakarta. *Urban Geography* 1-22.
[12] James, C. and Norquist, J. (2010). *Tearing down an expressway to restore a community.* http://www.nola.com/opinions/index.ssf/2010/08/tearing_down_an_expressway_to.html
[13] JUTPI Technical Team. (2011). *Transportation in Jabodetabek.* A presentation to the Coordinating Ministry of Economic Affairs. Jabodetabek Urban Transportation Policy Integration Project. April 12, 2011
[14] Leaf, Michael. (1994). The suburbanization of Jakarta: A concurrence of economics and ideology. *Third World Planning Review* 16(4): 341-356.
[15] Rukmana, Deden. (2014). Peripheral Pressures: Jakarta. *Archeology of the Periphery of Megacities.* Roger Connah (Ed.). Moscow: Strelka Press. Pp. 158-167
[16] Rukmana, Deden. (2010). Jakarta needs Metro to avoid traffic gridlock. *The Jakarta Post.* 24 October 2010. http://www.thejakartapost.com/news/2010/10/24/jakarta-needs-metro-avoid-traffic-gridlock.html
[17] Silver, Christopher. (2007). *Planning the megacity: Jakarta in the twentieth century*. London and New York: Routledge.
[18] The Jakarta Globe, October 11, 2013
[19] The Jakarta Post, 20 October 2010
[20] The Jakarta Post, 27 May 2011
[21] Winarso, Haryo and Tommy Firman. (2002). Residential land development in Jabotabek, Indonesia: triggering economic crisis? *Habitat International* 26: 487-506